Praise for
The Rocket Mass Heater Builder's Guide

Finally — a comprehensive book about rocket mass heaters! This should put an end to the freak shows of flaming death showing up on YouTube mislabeled as rocket mass heaters. And put an end to getting the same questions asked over and over again. We can start every answer with: "Have you read *The Rocket Mass Heater Builder's Guide?*"

— Paul Wheaton, founder of Permies.com online community

Eagerly awaited by homesteaders both urban and rural, appropriate tech folks, and do-it-yourselfers of all stripes, *The Rocket Mass Heater Builder's Guide* is a comprehensive, well-researched, and clearly illustrated manual on one of the most exciting home-heating concepts to come along in decades. The book is firmly grounded in the authors' many years of experience, will save every builder vast amounts of toil and frustration, and paves the way to a home-heating system that is clean burning, marvelously fuel efficient, and a pleasure to live with. Rich in detail without being overwhelming, this book is an essential addition to every homesteader's and builder's library.

— Toby Hemenway, author, *Gaia's Garden* and *The Permaculture City*

An advanced companion to *Rocket Mass Heaters: Superefficient Wood Stoves You can Build and Snuggle Up To*. The second, more comprehensive work on revolutionary wood heating; thorough, technically advanced, comprehensive, well thought-out by the experienced masters of rocket stoves.

— Ianto Evans, director and founder, Cob Cottage Company, inventor of Rocket Mass Heaters and Lorena Stoves

Drawing on the centuries-old tradition of masonry stoves and heaters, the "rocket" mass heater offers a modern, fuel-efficient way to heat a home with wood. This Guide represents more than ten years of collaboration and lessons learned from hands-on building and experimentation. The result is a frank, practical and detailed discussion of design and construction, pros and cons, and helpful anecdotal stories. While passionate about empowering the owner-builder, the authors also address code compliance and stress fire safety.

— Catherine Wanek, author/photographer, *The New Strawbale Home* and *The Hybrid House*, and co-author, *The Art of Natural Building*

In *The Rocket Mass Heater Guide*, Erica and Ernie Wisner have distilled the best, most tried-and-true design for efficient, low-cost, low-effort, carbon-neutral, regenerative home heating. Undoubtedly people will continue to experiment, but it will be hard to improve upon the wisdom of this book — what works, what doesn't and why.
— Albert Bates, author, *The Biochar Solution*

You will be happy to own this book for years and years to come! There are life-changing secrets inside of Erica and Ernie Wisner's most remarkable book, ways of understanding how to live better, more simply, and more beautifully while also being more snuggly and cozy that you ever thought possible!
— Mark Lakeman, co-founder, City Repair and design director, communitecture

Over the years, a handful of dedicated practitioners and authors have helped the natural building movement advance their craft based on solid engineering and experimentation. The authors are among that elite and vital group, and this book is a must-read for anyone who longs for practical, detailed knowledge on rocket mass stoves based on extensive experience.
— Dan Chiras, Ph.D., director, The Evergreen Institute and author, *The Natural House* and *The Homeowner's Guide to Renewable Energy*

Biomass energy is sensible when used super-efficiently and locally — which is the point of this book. Folks who live in cooler climates can benefit enormously from this simple, do-it-yourself technology.
— Richard Heinberg, Senior Fellow, Post Carbon Institute

The Rocket Mass Heater Builder's Guide provides a remarkable amount of detail for designing, building and using these elegant and efficient space heaters. In a world of ever-diminishing resources, growing populations and changing climates, safe, efficient and low-cost solutions to space-heating without fossil fuels are vitally needed. I have long planned to build a high mass rocket stove in my bioshelter/ greenhouse to replace our woodstove. After reading this book, I feel inspired and feel confident to do it.
— Darrell E. Frey, author, *Bioshelter Market Garden: A Permaculture Farm*

the rocket mass heater
BUILDER'S GUIDE

COMPLETE STEP-BY-STEP CONSTRUCTION, MAINTENANCE AND TROUBLESHOOTING

Erica Wisner & Ernie Wisner

Copyright © 2016 by Erica Wisner and Ernie Wisner.
All rights reserved.

Cover design by Diane McIntosh.
All photos and diagrams are by the authors unless otherwise noted. (See also image captions.) Cover diagrams, top and bottom row: by author. Center left: Greenhouse heater firebox, photo by owner; see color insert. Center right: Mediterranean heater with ceramic bangles, photo by Adi Segal — see Chapter 3.
Interior: Flames © AdobeStock_60101445; White brick © AdobeStock_69829337; Smoke background © AdobeStock_72849210; Adobe mud © AdobeStock_49425959

Printed in Canada. Third printing February 2018.

This book is intended to be educational and informative. It is not intended to serve as a guide. The author and publisher disclaim all responsibility for any liability, loss or risk that may be associated with the application of any of the contents of this book.

Funded by the Government of Canada | Financé par le gouvernement du Canada | Canada

Paperback ISBN: 0-978-86571-823-4
eISBN: 978-1-55092-616-3

Inquiries regarding requests to reprint all or part of *The Rocket Mass Heater Builder's Guide* should be addressed to New Society Publishers at the address below. To order directly from the publishers, please call toll-free (North America) 1-800-567-6772, or order online at www.newsociety.com

Any other inquiries can be directed by mail to:
New Society Publishers
P.O. Box 189, Gabriola Island, BC V0R 1X0, Canada
(250) 247-9737

New Society Publishers' mission is to publish books that contribute in fundamental ways to building an ecologically sustainable and just society, and to do so with the least possible impact on the environment, in a manner that models this vision. We are committed to doing this not just through education, but through action. The interior pages of our bound books are printed on Forest Stewardship Council®-registered acid-free paper that is **100% post-consumer recycled** (100% old growth forest-free), processed chlorine-free, and printed with vegetable-based, low-VOC inks, with covers produced using FSC®-registered stock. New Society also works to reduce its carbon footprint, and purchases carbon offsets based on an annual audit to ensure a carbon neutral footprint. For further information, or to browse our full list of books and purchase securely, visit our website at: www.newsociety.com

Library and Archives Canada Cataloguing in Publication

Wisner, Erica, 1977-, author
    The rocket mass heater builder's guide : complete step-by-step construction, maintenance and troubleshooting / Erica Wisner & Ernie Wisner.

Includes index.
Issued in print and electronic formats.
ISBN 978-0-86571-823-4 (paperback).--ISBN 978-1-55092-616-3 (ebook)

    1. Rocket mass heaters. 2. Rocket mass heaters--Design and construction. 3. Rocket mass heaters--Maintenance and repair. I. Wisner, Ernie, 1968-, author II. Title.

TH7438.W58 2016   697'.1   C2016-900792-8
                                          C2016-900793-6

Contents

Acknowledgments and Background vii

Chapter 1: Rocket Mass Heater Overview and Terms 1

Chapter 2: General Design Considerations 25

Chapter 3: Design Examples 39

Chapter 4: Step-by-Step Construction Example 59

Chapter 5: Operation and Maintenance 101

Chapter 6: Rules and Codes 137

FAQs 163

Appendix 1: Earthen Building 179

Appendix 2: Rocket Mass Heater Building Code (Portland, Oregon) 199

Appendix 3: Special Cases 205

Appendix 4: Home Heating Design Considerations 225

Appendix 5: Wood Heat Considerations 241

Appendix 6: Supplemental Practice Activities 257

Index 269

About the Authors 277

Acknowledgments and Background

Heating and cooling represents 30% to 60% of most household energy budgets. Rocket mass heaters offer a substantial savings on both heating and cooling in many climates. We hope that this book will prove useful as a small but practical step in addressing many key problems of our time: uncertain economies, deforestation and small fuels accumulation in our New World landscapes, the need for alternative energy and for more energy-efficient ways of meeting everyday needs, preserving clean air and water, and ultimately, making life on Earth more comfortable and pleasant for both humanity and the vast web of all life. Even if we can't know all that, we hope this book will at least give the reader hope for a more comfortable, safer, more affordable, and less toilsome future.

Rocket mass heaters would not exist without the pioneering work of Ianto Evans, documented in the Evans/Jackson book *Rocket Mass Heaters,* now in its third edition. On those foundations we continue to develop ecologically-sound, efficient, clean-burning heat, using locally available and nontoxic materials; and we continue to encourage owners to integrate these heaters into smart, efficient, compact, passive solar cottage designs.

Ernie Wisner joined this field as part of a dirty-hands research team under Ianto's banner at the Cob Cottage Company. Working in the US Navy, commercial fisheries, geophysical exploration, fire-fighting, and hazardous materials disposal, he saw and did some of the dirty jobs required by many common industries. He started looking for a field of work that wasn't self-defeating: something sustainable. Cob Cottage Company offered a dirt-cheap way to build permanent, healthy, and attractive dwellings, and a chance to practice appropriate/improvised technology for emergency preparedness and disaster relief.

During intensive research session in 2004–2006, Ianto, Ernie, and others built hundreds of test systems in an effort to diagnose reported problems. That work contributed to the first Evans/Jackson book collaboration, and to an ongoing grassroots development community including online forums hosted at proboards.com and permies.com.

Ernie wanted to document all the experimental data for inclusion in the Evans/Jackson book, but the team felt that too much technical information would overwhelm the average reader. They acknowledged the value of a technical manual, however, perhaps as a separate publication. Erica joined the efforts in 2006/7, with experience in writing, instructional illustration, and hands-on teaching. For the past ten years Ernie and Erica have been teaching workshops together, helping owner-builders to learn and build prototypes, and refining the design of these heaters for reliable performance in conventional buildings.

This book, then, is a hybrid of Erica's desire to present information in a step-by-step instructional manner, and Ernie's desire to share in-depth technical information. It offers more specifics and depth than the Evans/Jackson book, helping the reader avoid known problems, and clarifying our favorite methods for the trickier parts of construction, operation, and maintenance.

It is a privilege to be alive at this time when grassroots creative design collaborations can be undertaken with partners around the world. Builders on all seven continents have contributed examples, questions, problems, and discoveries to the general store of knowledge in this space. While we admit to some frustration when new builders replicate past problems instead of building on past success, the sheer quantity of examples available online helps prove out certain theories and rules of thumb much faster than could be done by any one team alone. We have learned a great deal from troubleshooting other builders' work as well as our own.

We are grateful to the friends, family, and colleagues who have supported us through the many years' process of developing and documenting this information.

We would not have been able to do this without the support of friends, family, and collaborators — particularly those who helped with contributions and work opportunities during the time immediately following Ernie's 2006 injury, and who helped make our work known through YouTube, permies.com, rocketstoves.com, richsoil.com, and magazines such as *Home Power* magazine, *Mother Earth News*, and *Permaculture* magazine.

Some colleagues and contributors are noted with their projects in Chapter 3 and Appendix 3, but many others have offered encouragement and insightful questions. Paul Wheaton and Kirk Mobert deserve our thanks as custodians of the permies.com and proboards.com forums.

Some noteworthy contributions to the field of clean heat include David Lyle's *The Book of Masonry Stoves;* Jim Buckley of Buckley Rumford Fireplaces and www.rumford.com, and Eric Moshier of Solid Rock Masonry, amazing masons who have been gracious enough to correspond about foundation work and chimney codes. The curators of the online Sweep's Library have expanded their shop website with a tremendous amount of down-to-earth public information at www.chimneysweeponline.com, and a similar public service is provided by the good folks at woodheat.org, engineeringtoolbox.com, builditsolar.com, and the HUD Heating Degree Days Index.

Thank you in general to those who've supported and documented the development of these heaters: Cob Cottage Company (www.cobcottage.com); Paul Wheaton (www.richsoil.com) and the Permies.com forums community; Calen Kennett (www.villagevideo.org); Bart Glumineau (www.studiomelies.com); Olivier Asselin (www.possiblemedia.org); Adi Segal (www.adisegal.com); Priscilla Smith; Arthur Held; Bryce Phelps; Eric Moshier (www.solidrockmasonry.com); Lasse Holmes; Max Edleson (www.FireSpeaking.com); Flemming Abrahamsson (www.fornyetenergi.dk); Peter van den Berg and Kirk Mobert (donkey32.proboards.com); Leslie Jackson (www.rocketstoves.com); and all the others who've helped to make this information available and accessible.

Thank you to all the owners, builders, workshop participants, and correspondents whose projects served as learning opportunities and proofs of concept. Most have opted for privacy over public acclaim, but we are especially appreciative of those who agreed to share their project details in Chapter 3.

We also wish to thank (and apologize to) those early clients and collaborators whose projects became unfortunate examples of what not to do, largely demolished now. We hope that the chimney-related failures in particular may be more easily understood and remedied with the information in these pages.

Colleagues who helped develop the Portland, OR, rocket mass heater code (Appendix 2) include Joshua Klyber, Bernhard Masterson, and the City of Portland's Alternative Technologies Advisory Committee.

For useful editorial and proofreading comments, we wish to thank Stephen, Anna, Rebecca, Catharine, Leslie, Lindsay, Rob, and all those who field-tested draft versions of this book. Thank you Linda, Rob, Sue, Greg, Maddy, and the whole publishing team from New Society and our overseas partners.

For personal support and generosity, and raising us along the way, we acknowledge and thank all of our families, friends, and collaborators. Any list would never be complete. So:

Thank You.

Chapter 1

Rocket Mass Heater Overview and Terms

What is a Rocket Mass Heater?

A ROCKET MASS HEATER is a heavy, slow-release radiant heater. It is designed primarily to heat people, secondarily to warm the areas in the line-of-sight around it. Modest tertiary functions include cooking, heating pots of water, and producing some warm air for distribution to other nearby areas.

A rocket mass heater is not a furnace or boiler; it should not be located in a hidden space, nor left to burn unattended.

The "rocket" in the name comes from a line of clean-burning cook stoves developed in the 1970s, using an insulated *heat riser*, to produce a very clean and efficient fire. "Mass" refers to the mass of masonry where it stores its heat. An RMH heater always includes an insulated, vertical, chimney-like heat riser as part of its clean-burning combustion chamber and a thermal mass to extract heat from its clean exhaust.

An RMH can warm a home using a fraction of the firewood required by other common heaters, such as woodstoves and boilers. The large, heat-storage mass these designs incorporate can heat multiple rooms from a central location, providing overnight heat without the danger and difficulty of an all-night fire. Rocket mass heaters are most effective in "line-of-sight" and by direct contact (built-in benches or sleeping platforms), although they do warm some air for heating distant rooms.

Not all builders of rocket mass heaters use the same design. This book shows our most popular "J-style" firebox design and a

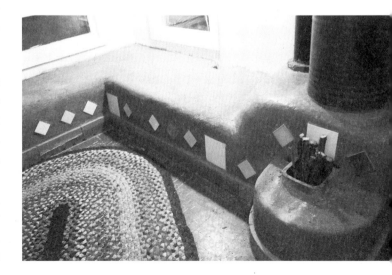

Annex 6" Heater photo © 2009 by Kacy Ritter.

heat-exchange channel that incorporates a metal pipe liner. Alternatives are described in Appendix 3, such as bigger batch fireboxes, alternative thermal mass styles, and other successful experiments.

Like other rocket stoves, RMHs use a narrow, well-insulated fire chamber to maintain a clean, hot fire. Their "whooshing" sound during full burn is pleasant (and much quieter than a propane heater). The typical fuel used is any local, dried firewood — from small branches to split cordwood. Unlike cooking rocket stoves, which have a reputation for requiring finely-split fuels, a working rocket mass heater can burn any log that fits in the feed without further processing.

Many high-end masonry heaters require custom ceramic parts, but an RMH can be built with common metal components, firebricks, and locally available insulation and thermal mass materials. Because the mass is horizontal with low, even weight distribution, they are relatively easy to lay out and install. Comfortable surface temperatures allow these horizontal mass heaters to be installed and used in a wide range of situations.

The original rocket mass heaters (see Ianto Evans and Leslie Jackson's, *Rocket Mass Heaters*, 2006 and 2014) were designed and built using local, earth-based, recycled, and reclaimed resources. Most examples in this book show standard components, such as firebrick and stovepipe. However, reclaimed and site-sourced materials are used by many owner-builders; Appendix 1 gives details on earthen masonry and workable alternative materials.

The purpose of this book is to allow you to build a properly working example on your first attempt — without replicating known pitfalls. For clarity, optional features and specialized designs have largely been omitted, though there is some discussion of alternative designs in Appendix 3. We encourage new builders to build a proven heater design first, and learn its ways before making any design modifications.

Parts and Functions

An RMH has two basic parts: the *combustion unit* and a *heat exchanger*. We will discuss the combustion unit first, and move on to the heat exchanger.

Anatomy of a Rocket Mass Heater

Parts and Functions

COMBUSTION UNIT

- Fuel loads vertically in the *fuel feed*. Air also enters here, controlled by a feed cover (not shown).
- Flames run sideways through the *burn tunnel*, then up the insulated *heat riser*.
- Together, the feed, burn tunnel, and heat riser make up the J-style *firebox*.
- Hot exhaust and fly ash fall down inside the downdraft *bell/barrel*, shedding radiant heat.
- The transition from this downdraft area into the heat-exchange ducts is called the *manifold* (see front cutaway view). ☛

Rocket Mass Heater Overview and Terms

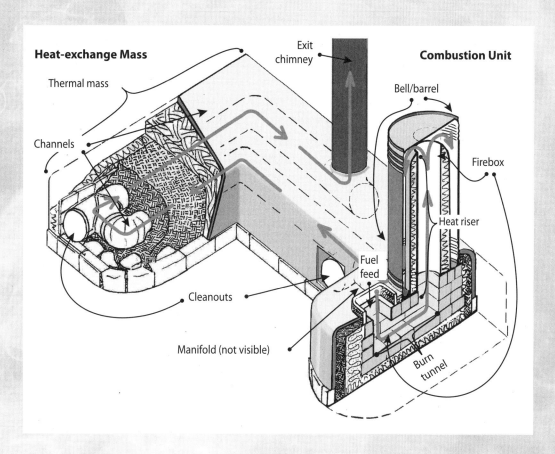

Heat-exchange Mass

- Exhaust flows through heat-exchange *channels*, transferring heat to dense masonry *thermal mass*.
- Exhaust exits the mass to the *exit chimney*.
- Capped *cleanouts* allow for annual inspection and cleaning.

Front cutaway view: Shows manifold (and exit chimney, dashed lines).

Combustion Unit

The combustion unit includes the fuel feed, burn tunnel, and heat riser (collectively the J-style firebox), and all the area under the bell (including the difficult-to-visualize shape under the barrel known as the manifold, which is the barrel-to-channel transition). The combustion unit experiences the hottest surface and internal temperatures. The right materials and correct proportions are critical for longevity, good draft performance, and clean combustion.

If a different type of combustion unit were substituted — one that did not burn as cleanly — there would be extreme danger of creosote buildup in the heat-exchange channels. In order to safely extract more heat from the fire's exhaust, that exhaust must be virtually smokeless. (When efficiency is the goal, a dirty fire is counter-productive anyway: smoke is wasted fuel.)

Combustion Unit Critical Functions

The role of the combustion unit is to create heat — and to do it cleanly. The flame path through the firebox interior must remain between 1000 and 2200°F (550 to 1200°C) for a clean burn. Consistently meeting this

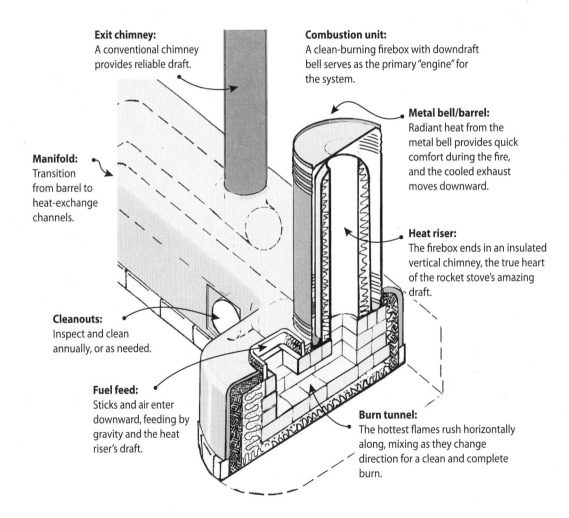

Exit chimney: A conventional chimney provides reliable draft.

Combustion unit: A clean-burning firebox with downdraft bell serves as the primary "engine" for the system.

Manifold: Transition from barrel to heat-exchange channels.

Metal bell/barrel: Radiant heat from the metal bell provides quick comfort during the fire, and the cooled exhaust moves downward.

Heat riser: The firebox ends in an insulated vertical chimney, the true heart of the rocket stove's amazing draft.

Cleanouts: Inspect and clean annually, or as needed.

Fuel feed: Sticks and air enter downward, feeding by gravity and the heat riser's draft.

Burn tunnel: The hottest flames rush horizontally along, mixing as they change direction for a clean and complete burn.

target despite the wide variety of natural fuels and weather conditions is not a trivial design problem; there may be damp willow and very dry oak or osage in the same woodpile. A firebox that reliably burns poor fuel above 900°F may occasionally exceed 2000°F with excellent fuel. (For reference, the temperature of flowing lava is often given as 1800°F; our clean fire can get hot enough to melt the Earth's crust.) To withstand a clean fire, we need special materials that are rated for thousand-plus-degree temperatures, known as refractory materials. Only clay-based brick or other suitable refractory materials should be used for the firebox interior.

Combustion Unit Components and Requirements

J-style Firebox

The fuel feed, burn tunnel, and heat riser together create a J-shaped path for fuel and flame. This narrow firebox or "J-tube" contains the flame path, similar to the firebox of a woodstove or other solid-fueled heaters.

Material: The firebox must be lined. Various materials can be used, depending on their "ratings," which indicate their ability to withstand heat. For a 6″ heater in a moderate climate, materials rated for 2000 to 2400°F (1100 to 1300°C) are adequate, such as perlite-clay and common clay building brick. For cold climates, extended burn, or anything larger than 8″ diameter, we prefer to line the firebox with materials rated for 2500°F (1350°C) or higher (firebrick, kiln brick, or rated refractory materials). Super-insulated fireboxes without a dense brick liner will reach hotter temperatures, and should be built of the highest-temperature materials available (high-end refractory insulation rated for 2800 or 3100°F (1530 to 1700°C), suitable for direct-flame exposure).

For proper draft performance, all the airflow channels in the system must provide the same *cross-sectional area* (CSA) — from fuel feed to exit chimney (with the exception of the barrel and manifold area, which may be larger). So, when building the firebox, you must create a consistent CSA. For example a system using 8″ diameter exit pipe has a CSA of about 52 square inches, so the brick firebox channels are built to 7″ by 7.5.″ See Chapter 6 for more on creating consistent CSAs.

Height: The hot, tall heat riser draws air and flames so strongly that it pulls air and flames *down* the relatively shorter, cooler fuel feed. This is the *thermal siphon effect,* making the combustion unit a *thermosiphon.* Like an ordinary siphon, this effect depends

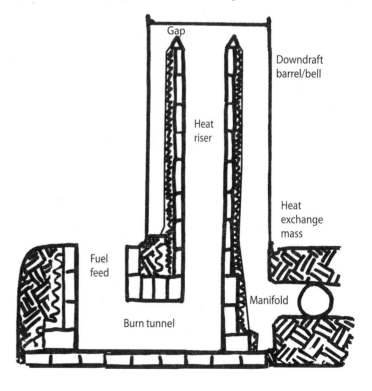

on differences in height and a complete absence of air leaks. Unlike a water siphon, though, the thermosiphon also depends on relative temperatures: hotter fluids move up, cooler ones move down. A hotter, taller heat riser works better. But a hotter or taller fuel feed can sabotage the draft.

Fuel Feed

This is the opening through which fuel and air enter the system; it is the first part of firebox. Fuel and cool air enter through the feed opening. The air strips heat from the top of the wood, reducing smoke-back and preheating the air. The fire burns hottest at the bottom of the feed. With good dry fuel, the sticks will burn all the way through at the bottom before the fire creeps up to their tops. They drop down as the bottom burns away, self-feeding and maintaining a bed of glowing charcoal for a steadier fire.

Fuel Feed Size: All fuel must fit within the fuel feed, so that the lid (not shown) can slide across for air control. Protruding sticks can cause smoke-back into the room. (See Chapter 5, Operation and Maintenance, for more details on fueling.)

The feed must be much shorter than the heat riser for the firebox to draw properly. Sixteen inches is a very practical height for the fuel feed (15″–16″ is a common standard for cord-wood cutters). A 12″ fuel feed is typically too short for locally available cordwood. The height of the fuel feed is related to the height of the heat riser. Our suggested 16″ fuel feed requires a 48″ heat riser, while a 24″ fuel feed would require a 72″ heat riser (6 feet tall).

Making a feed much larger than the system CSA requires a complete redesign of the firebox to prevent smoke in the room and creosote buildup in the chimney. See Appendix 3.

Durability: Normal fueling and cleaning can damage soft materials in the fuel feed and burn-tunnel areas. For this reason, we favor brick-lined fireboxes rather than pure refractory insulation.

Verticality: Natural wood fuels can self-feed downward for optimal air-fuel mixing and a steady burn rate, with minimal tending. The vertical feed, with its horizontal opening, also allows for an extremely cheap and effective fireplace door: two full-size firebricks can slide back and forth for air control or complete closure. (Half-size firebricks will get hot enough over a few hours' fire to toast bread, a nice bonus if you don't mind hot fingers.) With any fuel door capable of complete closure, it becomes the operator's job to remember that the fire needs air. The main burn usually runs best at quarter to half open (see Chapter 5, Operation and Maintenance).

If you need to shut things down before the fire is completely out, leave an air slot at least one quarter open (a 2″ opening) at all times. Closing the air off completely will not only starve the fire, it may allow smoke to back up into the room as the stove cools down too early, or even cause a flash-back fire (explosive smoke re-ignition) if the smoke-filled box is re-opened while hot. Indoor heating fires are best extinguished by letting the fuels burn down in a controlled way, not by campfire methods like water or smothering.

Burn Tunnel

This horizontal section of the firebox allows flames to pass from the fuel area to the heat riser. Coals often fall into this area to finish burning, and ash must be removed regularly during the heating season.

Some builders define the burn tunnel as only the area under the bridge (excluding the feed and heat riser areas). Our measurements are consistently given as the total lengths between the inside surfaces of the bricks, so the area we call the burn tunnel overlaps with the bottom of the fuel feed and heat riser.

Length: The minimum burn tunnel length is whatever is necessary to tunnel under the bell. With a smaller barrel or custom bell, the burn tunnel could be shorter without causing problems, but the bell must never overhang above the feed.

The feed bricks should be completely outside the bell, with at least ½″ of crack-resistant detailing between them, and space to remove and re-build the feed bricks if needed. Ideally, we prefer a total of 4″ to 5″ thickness of masonry all the way around the metal bell, especially if the masonry is providing both structural connection and airtight, crack-resistant detailing. However, this longer burn tunnel requires a taller or hotter heat riser, which may be impractical under a standard 8-foot ceiling. To take up less space in the room, we can use a metal liner extending down into the masonry for air sealing, and snuggle the brick feed right up close to the metal bell.

The maximum burn tunnel length is not more than half the heat riser height.

Shape: Brick channels, with their slightly rough surfaces, give a cleaner burn than a smooth or round firebox. (J-tubes were originally prototyped using metal stovepipe, however in addition to the problems of less-clean burn, metal also warps, cracks, and burns away rapidly when repeatedly exposed to high-temperature heating fires.) A square cross-sectional area is easiest to build and manage. When square is not practical with a given brick dimension, next best is slightly taller than square, making it easier to clean out the ash. Using 9″ firebrick to span the bridge creates a functional limit of about 7″ to 8″ burn tunnel width in any case, though this limit can be circumvented with other materials and methods.

Maintenance: The burn tunnel materials must withstand intense heat and periodic abrasion from ash-cleanup tools. Brick works well; softer refractory materials will degrade over time.

Bridge

The bridge is the ceiling of the burn tunnel, usually made with bricks or refractory slab.

Brick Replacement: The first brick in this bridge is part of both the fuel feed and the burn tunnel, where cool air and hot flame meet. This brick experiences intense thermal shock, strain, and damage from rough handling of fuel. Cracking is common. Allow for eventual replacement of this brick. We typically set this course and any feed courses that rest on it with a pure clay mortar, to allow for easy removal and resetting of all three bricks without rebuilding the entire firebox.

Many attempts have been made to extend the lifespan of this brick, but no alternative materials we've heard of have worked much better than a standard firebrick, which is also pretty easy to replace when needed. A full size firebrick will also provide many years of additional service despite a small hairline crack, while fancy solutions may fail in less-controlled ways and require earlier replacement.

HEAT RISER

This is the engine of the system. The heat riser is the insulated, vertical channel that draws flame upward and powers the rest of the firebox. Insulation helps hold the flames at high temperatures for complete combustion. A colleague compared the effect to an intentional chimney fire that burns the whole time, converting harmful smoke and creosote into clean, efficient heat.

Heat Riser Height: The heat riser must be substantially taller than the fuel feed opening (at least three times taller, measuring from the floor of the burn tunnel to the top edge of the riser). It should also be about twice as tall as the burn tunnel's length; this means the heat riser is always longer (taller) than the fuel feed and burn tunnel lengths put together.

Temperature: For good draft, you'll need to keep the heat riser hot. Surface temperatures can be 2500°F during a roaring fire. This means metal is not appropriate in the *interior* of the heat riser (or anywhere in the firebox); it will warp, crack, and eventually burn back down to ore. Ceramic materials such as firebrick, clay brick, or refractory ceramic components are the best choices for a heat riser's interior surfaces. Refractory insulation (2″ perlite or 1″ refractory blanket) keeps the inside hot and the outside cool; the exterior of this insulation usually stays below 600°F, so galvanized metal can be safely used for support around the outside of insulated heat risers.

Insulation: The heat riser must be completely insulated for reliable draft performance. Without insulation, the updraft and downdraft surfaces of the heat riser equalize about half an hour into the fire. The draft stalls, causing smoke to choke the fire (and the operator).

Insulation can be monolithic, such as a cast heat riser of insulating refractory materials, or it can be made from the light, foam-like insulating kiln bricks. Our preferred construction is a heat riser made of durable, dense firebrick, insulated with 1″ to 2″ of suitable refractory insulation such as perlite, ceramic-fiber blanket or board, or high-temperature rock wool. See Chapter 4, Step by Step Construction Example.

BELL/BARREL

The bell is also called the *downdraft bell, contraflow bell,* or the *barrel.* It is the container that redirects hot exhaust from the heat riser down into the mass. Typically made with a metal barrel or similar capped cylinder, the weathered metal surface provides quick radiant heat to the room and facilitates downward draft by slightly cooling the exhaust that flows through it.

"Bell" in masonry heating is a technical term for a chamber that traps rising hot gases to extract their heat. For proper functioning, its flow area must be *at least four times larger than the system CSA.* The gases stratify in this large slow-flowing space, with the hottest gases at the top, and the coolest gases sinking and escaping through the outlet near the bottom of the bell. In a well-proportioned heater, these cooler gases are still warm enough to rise upward in the exit chimney.

In the case of rocket mass heaters, the space under the barrel is not exactly a full-sized bell — it could also be considered as an oddly-shaped contraflow channel, with streams of hot gases flowing down the sides toward the heat storage bench.

Size: The practical bottom line: In the barrel and manifold areas, flow areas from twice to four times the system CSA seem to work fine. Larger may be fine too; we have not yet seen any such thing as "too much space" in this area. Too small a space creates problems: if there is less than 150% of the system CSA — or less than 1.5″ actual dimension between inner and outer surfaces — we see ash clogging and flow restrictions.

Complete Air Seal: Airtight, smoke-proof sealing is the most critical function of the barrel or bell. Any joints must be double-sealed with gasketing, clamps, or other heat-resistant sealers such as mortars or stove cements. (See Appendix 5 for maximum operating temperatures and appropriate ratings for high-temperature tapes, cements, etc.) Pinhole leaks in the barrel are not acceptable.

Radiant Heat: The barrel's primary purpose is not to trap hot exhaust, but to keep it moving toward the heat-exchange mass. For this reason, the barrel must *not* be insulated. Radiant heat loss from the barrel cools the exhaust gases so they can draft downward in this area and keep the system drafting correctly. See Appendix 6 for decorative and heat-shielding options around the barrel/bell.

Manifold

A manifold is a weirdly-shaped space where many pipes or channels connect, like a car engine's exhaust manifold. The manifold in an RMH is the transition area connecting the area under the barrel to the heat-exchange ducting opening(s), usually with an ash cleanout. So we have gases entering through the cylindrical slot/barrel-rim shape above, and exiting below to reach the heating channel(s) and a cleanout access opening.

The manifold may be constructed with an airtight brick plenum (box) with appropriate metal fittings mortared in place; as a hollow cob or adobe cavity sealed with earthen plasters; or by using a metal form such as a partial barrel or a slot-to-round heating vent, mortared or cobbed in place.

Manifold Air Seals: Exhaust leaks or air leaks in the manifold can interrupt the downward flow in the bell, causing stalled draft. Plan around any difficult-to-reach areas for a good seal and easy maintenance. For example, with brick work we often see novice builders having difficulty making good mortar seals in the corner behind the firebox, where access is difficult; with a fabricated steel part, the edges and fittings are the obvious weak points in the airtight seal. Rather than chasing leaks with your nose or a CO detector, consider allowing an extra ½″ of volume inside masonry manifolds to apply a crack-free earthen plaster inside. Better yet, consider finishing the exterior with a clay-based finish plaster, making any cracks very easy to detect and repair.

Manifold Area/Volume: The through-flow areas of the manifold need to be substantially larger than the system's cross-sectional area: at least 150% where the exhaust is changing direction, and about two to four times larger where it's merely flowing downward. Watch for bottlenecks created by fly ash buildup on any horizontal surface. A large ash pit below the heat-exchange ducting can help reduce blockages.

Manifold Cleanout Access: Provide access for annual ash cleaning. The cleanout needs, at minimum, to allow a hand or vacuum brush to remove any fly ash or obstructions between the heat-exchange channels and the barrel downdraft areas. If there is an ash pit, the cleanout should be able to reach it too.

We always keep this cleaning access in mind as we build — you want to be able to reach into the manifold and pull your hand back out intact, and the difference between permanent structure and ash (sometimes hardened by years of neglect) should be clear even to gloved fingertips.

With adobe or cob manifolds, we sometimes build in a tile floor, or use a good hard plaster on the interior. Perlite insulation is easily mistaken for ash, and needs some plaster or tile reinforcement. With blanket insulation, we always provide a metal cage strong enough to hold the blanket against a vacuum nozzle's pull, and sometimes plaster this as well.

Cleanout access may be:

- a capped T in the first section of exhaust pipe
- a capped pipe or tile cleanout in the manifold itself (mortared into masonry manifolds)
- a removable barrel or barrel-lid, air-sealed with fiberglass woodstove gasket. (But lifting an entire barrel carefully over the heat riser every year will make you wish for an easier cleanout.)

Heat Exchanger

The heat exchanger includes the system of channels and masonry that capture exhaust heat: the heat-exchange ducting, thermal core, and casing. The most popular and efficient layouts are horizontal, creating a bench, bed, or heated floor.

Heat Exchanger Critical Functions

Thermal mass storage: To function as described, an RMH needs a large mass of dense material to store heat. The materials used vary. Metal, brick, clay, stone, or even water can all be used as heat storage mass. (Water, though, is a lively and weird heat-storage material with its own special challenges; see Appendix 3.)

The original rocket mass heater designs all used a clay-based earthen concrete called cob — and they used a lot of it. A typical rocket mass heater's masonry may weigh 3–6 tons (which is distributed on the footings at about 120–200 lbs/sf). The transportation of such heavy materials is a large part of their cost and environmental impact, which is why we prefer to use materials from as close to the project site as possible (whether it's stone, cob, adobe, tamped mineral soils, or recycled concrete rubble).

Heat storage capacity for a given mass of material is measured by the material's heat capacity, weight, conductivity, and degree of temperature difference. It's possible to store the same amount of heat in less mass if you can get the mass hotter ... but the temperature of the small, hot mass will also drop more quickly, delivering that heat in a shorter time.

Airtight: The channels that carry exhaust through the thermal mass should be double-sealed to prevent exhaust leaks. The primary seal is created when building the channels from metal pipes, clay pipe or chimney liners, or mortared brick. A

Insulation Note:

Some people think insulation "stores heat" or "keeps things hotter." But, as it goes with a Thermos, insulation can also keep things colder. What's going on?

Insulation actually blocks heat *transfer*. The insulation itself typically doesn't hold much heat, being mostly air, so it prevents heat from reaching anything on the other side. Thus you typically *don't* want to put insulating materials between the exhaust pipes and thermal mass, or within the thermal mass; this would stop your heat from reaching its storage (and you!).

Insulating materials to watch out for in the thermal mass core include hollow straw, and any material with empty air pockets (loose sand, gravel, sloppy filling around the pipes, etc.). Insulation is sometimes desirable around the *outsides* of the mass, to prevent heat transfer (for example along an exterior wall), or as removable cushions that trap and control the stored heat for later use (for sleeping cool, then tucking toes under the cushions next morning). A small amount of straw or fiber in the outer casing can be useful for other reasons.

---

secondary seal is provided by wet-formed masonry in direct contact with these lined channels. If there is no wet-formed masonry around the pipe channels, then a reliable air seal must be provided some other way (for example, welding the pipes, and installing CO detectors).

Efficient Heating: Conductive heating is highly efficient. A full-body heating pad or heated seat keeps occupants deliciously warm, even at room temperatures cooler than normal. Radiant floors or low heated benches reduce stratification of room air; instead of hot air rising up to the ceiling and looking for ways to escape the house, more warmth stays down where it's most useful, resulting in greater overall comfort and less wasted heat.

Safety (structural and fire/exhaust): The thermal mass bench must be airtight to prevent smoke leaks and control any condensation/drainage. It must remain structurally sound through decades of use, possibly including earthquakes, floods, and changes of family life. Structural integrity must carry through both the core and casing.

Heat Exchanger Components and Requirements

CHANNELS

Lined channels carry the warm exhaust through the mass, transferring heat through the channel walls into the storage mass. Using standard metal pipes makes it easier to lay out and set slopes for these channels, but channels can also be built with any material that will tolerate the heat, damp and corrosive exhaust, and occasional abrasion from cleaning: clay chimney liner, clay drain pipe, or smoothly-mortared brick.

Consistent Cross-sectional Area (CSA): From manifold to chimney, all the ducts are the same cross-sectional area. Ideally, the areas connect smoothly, without corrugations or obstacles to induce turbulence, and the channels are shaped as circles, rounded squares, or other similar shapes that allow excellent through-flow.

Heat-exchange Duct Length: The heat-exchange channels extract and store useful heat, but they also represent a load on the system's draft (through friction and cooling). A 6" or 8" diameter system typically has 20 to 40 feet of ducting length, and at least one 180-degree turn. There is no practical minimum length, but a shorter system will store less heat and may exhibit over-drafting (drawing too much air for the fire). See Chapter 6 for specific rules of thumb regarding length.

Orientation: Roughly horizontal layouts are the norm; they distribute weight loads and offer maximum comfort seating. We slope the pipes upward slightly, ¼" over 8 feet or so, or about 1" over the whole run in a level bench. This slope allows for upward gas flow, downward drainage of condensation water to a cleanout, and provides even heating on the surface by locating the hottest pipes lower and the cooler pipes closer to the surface.

Where a vertical configuration is needed to fit the space, channels should run roughly horizontal or with a slight rise between turns, just as in a horizontal bench. Avoid zigzagging up-down runs because they trap hot and cold air, causing substantially more draft resistance and inconsistent heating.

Smooth Inner Lining: Only smooth, easily-cleaned materials should be used to line the heating channels. Rough or corrugated textures can create drag equivalent to tens of additional feet of length *per foot*, seriously affecting the system draft. Avoid corrugated pipe, wrinkly corrugated elbows, wrinkly flexible ducting, or chimney-liner sections with spiral corrugations. Adjustable elbows that rotate are smooth enough to work well, and T's for cleanouts seem to have little effect on drag. Large voids, such as cleanout extensions through an exterior wall, may affect draft.

Maintenance: As in the combustion unit, all linings should be durable enough to tolerate manipulation of a brush, drain snake, or Shop-Vac hose.

Capped Cleanouts: Cleanouts need to allow maintenance access to every part of the system. Critical cleanouts are located at each 180-degree bend and at the bottom of each vertical drop such as the barrel/manifold and the exit chimney. Any series of 90-degree bends may also need additional cleanouts to give access to each pipe. Imagine using a periscope mirror and flashlight to try and find an asphyxiated squirrel in there, and make it easy.

Fresh Air Separation: If any channels are built to convey outside air or room air, these must be completely separate from the exhaust ducting channels. Be extra careful in any areas where an exhaust leak would be hard to inspect, such as the wall side or interior of the heater.

The exit chimney opening should likewise be located outdoors, above the roof ridge, a suitable distance from any nearby air intakes, doors, or windows.

THERMAL MASS

Thermal mass refers to all the masonry surrounding the heat-exchange ducting, including the *core* and the *casing*.

Anatomy of a Rocket Mass Heater

Heat exchanger: Channels through mass extract and store heat from exhaust.

Heat-exchange channels: Metal pipes allow for easier construction and maintenance, airtight seals, and good heat transfer.

Thermal mass: Dense masonry (often earth or local fill) slowly heats and cools, preserving warmth for hours or days.

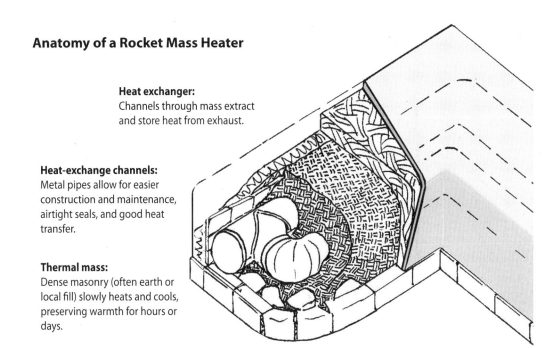

Density/Heat Storage: Thermal mass is dense so it can store more heat over hours or days. Light, foamy, or airy materials block heat transfer and have less heat-storage capacity.

Reactivity: Thermal mass materials should be nontoxic, chemically resistant or inert. They should not corrode or degrade if exposed to exhaust leaks or minor water damage, and be completely unaffected by moderate heat. The hottest areas are immediately around the firebox, manifold, and first run of heat-exchange ducting. Temperatures of the interior mass in these areas may exceed 500°F during operation, with surface temperatures up to 250°F.

Conduction, Not Insulation: Airy materials such as perlite, pumice, straw, and dry sand offer insulation, not heat storage. Insulation may be useful between thermal mass and exterior walls, but not within the mass itself. Thermal mass should be relatively dense and conductive, the opposite of insulation. If airy or light materials are included in the mass (either to reduce weight, or in a misguided attempt to save the heat), they will reduce the heat transfer and heat-storage capability of the mass.

Core, Thermal Cob

This is the dense material that surrounds the channels of the heat exchanger. Usually it is earthen masonry or fireclay-mortared unit masonry. It is the permanent structure around the ducting, and provides the immediate heat transfer and most critical heat storage mass.

Secondary Seal: The core of the heater provides the secondary seal and heat-transfer functions around the heat-exchange ducting. The ducting may last for decades, but good masonry detailing here can last for centuries. If you notice any voids, cracks, or crumbling areas in the core materials as they dry, these cracks should be patched, or the faulty material should be removed and replaced before finishing the casing.

Heat Conduction: Core materials should be fairly conductive — comparable to brick or earthen masonry, which transmits heat at about 1″ per hour. If using an infill technique (such as described for greenhouse heaters or some portable designs), try to find mixed, small aggregates such as ¾″-minus rock-crusher gravel or quarry fines, masonry sand, or a suitably gravel-rich natural mineral soil. Avoid sifted sand, beach sand, perlite, or other uniform and air-trapping materials.

No matter the size of the particles, mixed sizes and sharp edges are easier to pack and stack, and dense packing gives better heat storage. (Four loose buckets of good masonry sand plus 1 bucket of clay might make 2.5 packed buckets of fill material — it gets a lot denser as you mix.) A mess of identical round sand grains or river rocks are hard to stack, and it will take a lot of clay, mortar, or fill materials to fill those air pockets or they will block heat transfer.

Infill: The core of the bench can incorporate any available masonry rubble including concrete chunks, rock, gravel, even pottery shards or broken dishes. Avoid dangerous materials such as broken glass, vacuum tubes (strange but true, people have buried old TVs in cob projects — repair crews beware!), contaminated factory wastes, or heat-sensitive/flammable materials.

Casing

Also referred to as cladding or veneers, the casing is the final, outer layers of the heat-exchange mass and the supporting masonry around the combustion unit. Typical casing materials include earthen plaster, stone, brick, tile, and some other materials that offer a durable and attractive finish. Some builders have used wood trim, but only in well-tested locations away from high-heat areas.

The casing can be built first and then filled (for example a brick box, with the ducting just 1″ from the inside of the brick), built in courses tied into the heater's core (like rough stonework), or applied later as a finish layer (such as plaster or tile). Materials that could trap moisture are best applied after the core is completely dry. If the heater has been thoroughly test-fired while drying, these decorative materials can hide any patched cracks from the drying process or heat-expansion.

Durability: The casing materials should stand up to ordinary wear-and-tear. *Normal* use includes people sitting, sleeping, dining, dropping things, wrestling on the sofa, and climbing all over it like monkeys. Common variations might include sheltering and watering houseplants, using heat to quicken home brewing, quick step-ladder function to water hanging plants or change an overhead light … possibly all simultaneously. Crayon-resistance is optional.

Repair or Refinish: Some materials are easier to repair than others (see Appendix 1). It is helpful to set aside color-matched plaster for later patching.

The best masonry heaters are designed for eventual piece-by-piece inspection and repair; they last longer because they can be fixed or remodeled as needed. Some modern masonry materials bond too strongly for practical repairs; instead of repointing soft mortar or grout, a crack in these modern materials may require complete replacement of broken bricks and tiles.

Moisture Protection: The heater should be well protected from damp, including ventilation and drainage of the heater base to dry out construction moisture. But don't try to waterproof the entire outer surface: trapped internal moisture can cause blistering or separation of the casing layer, or unsightly mineral blooms. Breathable, heat-tolerant finishes can be used to create a water-resistant, durable surface. Waxes, oil paints, and volatile oils generally are not suitable for heated surfaces.

See Appendix 1 for more detail on natural plasters and other compatible finishes.

Heat Tolerance: Operating temperatures for the heat-exchange mass itself tend to range between 70–100°F for room heaters (20–40°C), or up to 150°F (65°C) for larger house heaters. Combustion unit temperatures and some high-output designs may have higher surface temperatures (up to 250°F/120°C).

Natural wood trim can be used on areas that never exceed 150°F (65°C) in the heaviest use. Natural fibers can be included in mineral plasters (clay, lime, gypsum plasters) without making the plaster combustible.

Fabric Choices: Natural-fiber fabrics and battings that tolerate warm or hot settings on an iron or tumble-dryer (roughly 300 to 400°F/150–200°C) are suitable for most heat-exchange benches or beds. Remember that cushions are insulation; heat trapped underneath cushions can raise the surface temperature of the masonry higher than if it were bare. Check to make sure all surface temperatures in contact with wood, fabric, cushions, or other combustible materials remain below 150°F (65°C) — not painfully hot to touch. An isolated hot spot can be covered with thicker masonry (tile, plaster, or a sculptural element such as an arm-rest over the hottest areas adjacent to the barrel). Or you could add another inch or two of tile or plaster to the whole bench, to absorb some of the excess heat.

Exit Chimney

The exit chimney is a sealed pipe or flue that conveys exhaust out of the building. Proper chimney design helps ensure good draft for the heater, despite varying wind and weather. Most installations use a standard manufactured chimney, a low-temperature appliance exhaust, or a lined masonry chimney.

Draft: Vertical, warm chimneys provide draft that keeps the whole system operating smoothly, and provides a slight under-pressure to prevent smoke escaping any pinhole leaks. Unconventional chimneys often cause draft problems.

Structure: Chimneys should be physically self-supporting independent of the heater

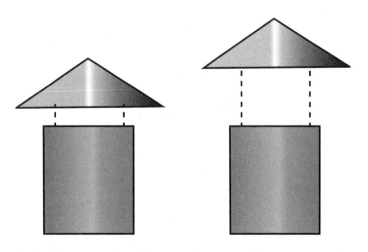

A taller chimney opening allows enough screen surface area for 100% flow.

itself (in case of building problems, refits, or repairs).

Maintenance: Chimneys should have a cleanout near the bottom of the vertical section, an appropriate rain cap and screen (with through-flow area of at least 100% of system area).

Wind and Weather: The simplest and most reliable options are generally dictated by local building codes and best practices for combustion appliance exhaust and woodstove chimneys. Wind gusts, snow loads, ice dams, rain, prevailing winds, seismic loads, differential frost heave or settling of buildings, and variable outdoor temperatures create special problems for chimney design.

The easiest place to make a chimney exit weatherproof is actually near the ridge top, where there is less rain and snow coming down the roof toward the installation, and it's easier to lap the chimney flashing over any type of roofing. This also allows the building's rafters or trusses to support most of the chimney, while raising it to the proper height above the gusting chaos of building-related wind eddies.

Temperature: Woodstoves must exhaust above 350°F by law in the US (about 175°C, but local laws vary). Fireplaces and woodstoves often exhaust as hot as 600–800°F (300 to 450°C). Masonry heaters generally exhaust between 200–400°F (90 to 200°C). Rocket mass heaters can draft substantially lower, as low as 60°F (15°C) in some odd cases outside the realm of normal building practice.

Some efficiency-minded colleagues glory in seeking out the absolute minimum exhaust temperature. A few of Evan's original heaters ran fog-like exhaust well below its dew point at 60 or 70°F (15 to 20°C). This is very exciting for efficiency fans, as almost no heat is being lost outside the building. Unfortunately, this exhaust emerges as a cool, dense fog that will not rise in a conventional chimney, and it lacks the force to overcome even the slightest wind pressure. These ultra-low-temperature exhausts must exit nearly horizontal or downward, and are difficult to protect from normal wind and household pressure conditions. Too-low exhaust temperatures create draft problems in almost all conventional buildings, and are especially unreliable in winter storm conditions (just the time when an electrical exhaust fan is also vulnerable to power failure).

The only conditions in which a balky heater with a tendency to run backwards would be acceptable are conditions where winter heating is not critical for survival — and in those cases, passive solar heat collection and a smaller, seldom-used wood heater would generally be a more practical and more energy-efficient solution.

For reliable draft and efficiency, we generally target a surface temperature of 100–150°F (40–65°C) on the exposed exit chimney pipe during test-firing. Average exhaust temperatures inside the pipe are likely close to 200°F/90°C, the low end of masonry heater exhaust).

For further discussion of best practices, see Chapter 4, Step by Step Construction Example and Appendix 3, Alternative Chimneys.

Major Design Considerations

Here are some general "rules of thumb" to follow when installing a rocket mass heater. These rules are covered in detail in Chapter

6. However, a few key points may be important at this stage.

Cross-sectional Area (CSA): The basic rocket mass heater designs use the same cross-sectional area for all channels, with the exception of the downdraft barrel and manifold (which may be larger).

Combustion Area Proportions: The heat riser height drives the whole system's draft. The heat riser should be at least three times taller than the fuel feed, and at least twice as tall as the burn tunnel length.

As an example, an 8″ heater that meets these proportions might have a 16″ fuel feed, a 24″ burn tunnel, and a heat riser 48″ tall. The heat riser could be taller (up to about 6 feet) but not shorter.

Length of Heat-exchange Pipes: The goal is to extract as much heat as practical, without making the system too cold for comfort or too difficult to start. Climate, heat load, and available space all factor into the desired size of mass. The mass and pipe length are not the only load on the system draft: the number of turns, rough texture of pipe, or other obstacles to smooth flow will greatly reduce the draft and a shorter pipe may be needed to compensate.

A typical heat-exchange bench runs about 20 to 35 feet of heat-exchange channels, with one or two 180-degree turns, for a cool-climate exhaust that runs about 100 to 150°F (pipe exterior temperature).

For hotter climates or trickier operating conditions, a shorter bench and a hotter exhaust (200–250°F) may give more reliable draft. A heater with horizontal channels between 12 and 20 feet in length will exhaust in this hotter temperature range. The sacrifice in fuel efficiency is only about 10%.

So if you have a variable heating season (such as not heating if you're gone on vacation), the climate is hot or variable, you are not able to store two years' supply of dry wood, or you want to be able to operate the heater in warmer weather for saunas or visitors, consider a shorter bench run.

A poor exit chimney (cold, less than half the building height) is a tremendous obstacle to good draft. The condensation of water in the exhaust can cause chimney stalls, which helps explain why cold, exposed outdoor chimneys often draft poorly, even if they never become actually colder than outdoor air.

Detailed rules of thumb, building codes, and workable proportions are given in Chapter 6: Rules and Codes.

Is a Rocket Mass Heater Right for Me?

What Rocket Mass Heaters Are and Aren't

Rocket mass heaters are large space heaters, typically installed in a central part of an occupied home. In the right location, they are extremely efficient. In the wrong setting, they are a waste of effort.

- They provide steady, ongoing heat. They work best in a continuously-occupied home (as opposed to a weekend cabin or chapel).
- Rocket mass heaters are not automated central heating. They do not replace a furnace or boiler; they are not designed to run unattended in a basement. Instead, they are a form of gentle radiant heater, and they perform best when centrally located in the occupied space. Many owners

find that their rocket mass heater substantially offsets the furnace bills, and heats several adjacent rooms nicely, but the heat may not evenly reach the farthest rooms.
- Human Operated: For reliable heat in case of grid power failures, our designs are fully manual, no electric automation. This means a human operator must supervise the fire.
- Unusual: Local officials may have no idea how to permit or inspect one, and insurance companies may need to be approached (and educated) about coverage. There is a learning curve for operating a downdraft heater, as the fire is upside down compared to most woodstoves or fireplaces.
- Site-specific design and installation implies local knowledge and intelligent choices. Unique designs are more common than any standard model. When building in a new situation, it's possible that under those new conditions, an identical heater will not function as described here.
- Finally, the designs in this book are intended to burn only natural wood fuel such as brush, branches, poles, small logs, split cordwood, lumber scraps, and small amounts of other natural fuels such as paper tinder. For examples of alternative fuels designs, see Appendix 3.

Considerations

Benefits

An RMH provides efficient heating that matches the needs of many American households, especially for a primary residence. They offer safe warmth overnight; efficient use of local, renewable resources; and affordable, DIY-friendly construction and maintenance. Rocket mass heaters do not duplicate the (inefficient) functions of a furnace or boiler. Part of the way they save energy is by heating people instead of space. (See in Chapter 2, "Heat Transfer" for details.)

To achieve the best possible results, these systems rely on locally available materials, knowledge, and fuels (ideally waste or thinned fuels, not standing timber). They offer simple and satisfying operation and maintenance, and low construction costs (competitive with a new woodstove). When properly constructed they will produce a clean fire (no creosote or wasted fuel).

Most masonry heater designs share these goals of safe warmth and fuel efficiency, and many achieve similar clean burning and heat storage. However, most masonry heater designs of comparable capacity are more expensive to build due to cast-ceramic parts, metal and glass doors, fancy veneers, and other popular design elements. Each style of masonry heater has its own special design rules and tolerances. (See *The Book of Masonry Stoves,* by David Lyle.)

The designs we use were developed within North American culture, specifically within a grassroots community of self-sufficiency enthusiasts. Rocket mass heaters are attractive to this community because they offer grid-independent or off-grid heating, and they are good alternatives for single-family or single-occupant homes whose owners are busy and have limited hours to tend the fire. They also offer a quick return on investment. This is important in our context because, on average, Americans move their households every five years.

We can hope, but we do not confidently expect more than five years' return on any investment in home improvements.

Some Pros and Cons

Convenience at modest cost comes with some downsides. Here is a quick review of some pros and cons of RMH heaters:

Pro: Quick payback. Fuel savings will often repay installation costs in less than four years — often less than one year for resourceful scroungers replacing inefficient heaters.
Con: The most effective, cheap materials are not always the prettiest, and there can be conflicts with local building codes.

Pro: Rocket mass heaters are routinely built by novice masons and owner-builders. Expert help that might be available could include interested heater masons, natural building contractors, and heating engineers; such professionals may be able to help with projects challenges, including local legal questions.
Con: Work done by non-professionals may be less acceptable to local decision makers. Owner-builders often have more legal freedom than landlords, mortgage-holders, or renters.

Pro: Materials are readily available. Commercial, reclaimed, or recycled parts and local mineral materials may be used for mass.
Con: Using reclaimed or site-sourced materials may not be permitted under building codes; upgrading to new, modern materials may involve unexpected costs. Using site-sourced materials effectively takes some skill and practice.

Pro: Heat output is both steady and responsive. A traditional masonry heater requires 12- to 24-hour advance operation. A woodstove heats instantly, but cools just as quickly. Rocket mass heaters provide quick radiant heat from the metal bell, and steady stored heat from the bench: best of both worlds.
Con: The metal bell's radiant heat can overpower a small room during extended firings; all-masonry heaters may heat more evenly. The bench takes time to heat up after an absence, and will also store heat even if not wanted (such as if you fire it up to cook in summer).

Pro: Rocket mass heaters are not dependent on power grids or imported sources. Heater self-regulates to a degree, but rewards conscientious operation.
Con: You may need a house-sitter during winter vacations to light the stove two or three times per week, if you need to keep plants and plumbing intact in cold climates. If your sitters can't be trusted or trained, you'll rely on other options like an old but working furnace, heat-tape for the pipes, or using frost-free plumbing design.

Drawbacks

THEY TAKE UP SPACE

Rocket mass heaters do fit into a surprising variety of situations, having simpler foundation requirements, more flexible design configurations, and lower heights and clearances than many other masonry heaters. But they still take up a substantial amount of space.

Most North Americans use hidden furnaces or a compact (if uncomfortable) woodstove. Modern homes may not have an ideal space for a mass heater.

Traditional masonry heaters are often built first, then the house is built around them. A well-integrated masonry heater can serve as a dividing wall to warm three or more rooms, and other nifty features can be added. Retrofitting these features into an existing space is trickier.

The optimal place for effective heating is to locate the heater wherever you sit the most; you can replace the woodstove plus your favorite sofa. If that's not possible, consider a new dividing wall or section of floor (with the heater mass sunk into the crawl space). The most effective heating location is therefore, unfortunately, the one that will most disturb a beloved hangout spot.

Change is not fun, and living in a construction zone is not fun, so consider domestic tranquility in your planning. It may be less disruptive to build the heater into a new addition on the house, even if the resulting heat is less efficient than with a centrally placed heater. Or, your rocket remodel could redefine the favorite hangout spot, turning an under-used space or bump-out into the hot new hangout spot.

Non-standard Technology

In this era of global appliance manufacturing, site-built masonry heaters may be rare in your area. Local officials, co-owners, mortgage lenders, and others may need time to get used to the idea.

Do you need approval for your project? Mortgage or insurance contracts may impose strict penalties if you build your own heater without official approval. Some owners have had success getting local permits, or discovering applicable exemptions; others have run up against weird local regulations or skeptical building officials.

See Chapter 6 and Appendix 2 for more about local codes.

Learning Curves Can Be Steep

While the vast majority of owners have little trouble adapting to their new heaters, it may be a different story for guests or new tenants. Contractors unfamiliar with the design may introduce almost as many problems as they solve. Well-meaning visitors can have a great deal of trouble guessing how to correctly operate the heater.

If possible, visit others' heaters during the heating season, or build a prototype in an outbuilding. Nothing convinces the household holdouts like the experience of sitting on the warm bench in chilly weather.

Loading Takes Time

A rocket heater can only burn so much fuel in a day, or in the available hours that people can watch the fire. But the fire should not burn unattended — though the operator can easily combine fire tending with other routine activities nearby. Keep an ear or eye on the heater from any nearby room, refueling or adjusting the fire every hour or two. Hours of operation vary with heating load.

In modest, well-insulated homes (up to about 1500 square feet), an 8″ system might burn:

- 1 to 2 hours per day in cool/mild weather (30–60°F/0–15°C)
- 4 to 6 hours per day in chilly weather (0 to 30°F/-15 to 0°C)
- 8 to 10 hours per day in sub-zero cold (-30 to 0°F/-35 to -15°C).

Larger homes, leaky or thin-walled buildings, or an extreme climate will require more fuel, and thus more time, to heat. In some cases a rocket mass heater may not be enough to offset all building losses; you may need a backup heat source for really cold weather, or you may have to make some home heating efficiency improvements.

Multiple heaters are not very convenient to run unless they can be automated (or you have more operators to share the work than heaters).

If you want to heat a large home on wood alone, with minimal operator hours per day (not counting wood preparation), consider large masonry heater designs that have larger batch-burn fireboxes, or automated central heating (furnaces, boilers). They will not be as fuel efficient, but heating large spaces with small numbers of people inside is not an energy-efficient lifestyle to begin with.

For all spaces, consider ways to reduce the heating loads. Can you capture or conserve more heat? Can you cluster the heat-loving elements into a smaller space? Can you improve insulation, or close off unused closets and rooms to protect the warm core?

We have known people to keep their original furnace or woodstove, and use the mass heater to cut their overall fuel bills (often to one quarter or less of the original costs) while retaining the old heater for backup.

In general, rocket mass heaters have been most popular for small, efficient homes, in temperate to sub-arctic climates. Arctic and industrial-scale designs are beyond the scope of this book, except as noted in Appendix 3.

Firewood Takes Work

Those moving from automated heaters (furnace, heat pump, baseboard heaters) to wood heat may not realize what's involved in securing and storing dry firewood.

Dry fuel is not found in nature, nor is it found on damp ground under a tarp. Like dehydrated fruit or dry laundry, dry firewood is a product of intelligent effort. The tool for producing dry fuel wood is a wood shed. A good, dry woodshed repays the investment many times over with lower fuel costs, less work, brighter fires, safer chimneys, and easier (and cheaper) chimney maintenance.

Wood drying requirements vary by climate, but plan to store a minimum of two years' supply: half for use this year, half drying for next year (or in case of an unusually hard winter). Some varieties of wood take longer to dry.

If you currently have an almost-big-enough woodshed, a rocket heater's fuel efficiency may relieve you of the need to build a larger one. If you currently have no storehouse capable of producing or storing dry fuel, your budget for a wood-burning heater MUST include dry fuel storage. For details, see Chapter 5, and our companion booklet *The Art of Fire*.

Limited Flame-viewing

One of the customary perks of heating with wood is the pleasure of watching the flames. The efficient J-style firebox is intellectually amazing, but hard to watch from across the room. The operator gets the best view, though the flickering display on the ceiling can also be fascinating. Mirrors or reflective surfaces near the feed (such as a band of shiny metal trim on the barrel) can offer a reflected glow.

Ceramic glass can be used with care in some areas of the firebox; however, the glass is not as insulating as an 8″ masonry and insulation sandwich, and the clean burn seems to suffer. Getting a glass-fronted door suitable to anchor in masonry (for example, from traditional oven suppliers) can add $500 or more to the project cost.

If fire-watching is a serious hobby in your household, consider a separate "fire TV" for occasional use, such as a glass-door woodstove, Rumford fireplace, or rack of candles. These low-mass space heaters can be perfect for a parlor, guest suite, studio, or formal dining room that is not used every day.

Some Spaces and Uses Are Unsuitable

Occasional-use Spaces

Thermal mass is slow to heat up and slow to cool down. It takes time and energy to shift the heater back from cold-slab mode (lovely in summer's heat) to warm-cuddly mode. If the heater mass is colder than outside air, the draft can be reluctant to start.

If you have a vacation cabin where you spend the occasional long weekend, it may not be the best place for a large masonry heater. You spend Friday night and part of Saturday in a chilly cabin bringing the mass up to temperature, enjoy the warmth briefly on Sunday morning, only to head back home on Sunday afternoon and waste the remaining stored heat.

While a rocket mass heater's radiant barrel does offer more responsive heat than most masonry heaters, making Friday night cozier, the heat in the mass is still wasted on Monday and Tuesday instead of being available for your enjoyment.

That ski cabin might be just as comfortable with a simple radiant heat source like a Rumford fireplace or woodstove. Even an inefficient woodstove or space heater may use less energy over time if heat is only needed in short bursts. In general, for heaters that will be warmed up on demand for only a few hours, we prefer a thinner mass (2" to 4"); or a lower-cost option such as a wall of brick behind a woodstove.

The same problem occurs in churches, meeting halls, retreats, and guest facilities: preheating the mass takes precious custodial time and energy, and there is the same problem of wasted heat after the group leaves the space.

Passive solar designs and climate-appropriate designs, including insulation for harsh climates, are a good investment no matter what the building's used for.

Comparing Other Heating Options

When we suggest that a rocket mass heater is not a good fit for a particular situation or goals, we are often asked about other heating options. The Heating Situation graph compares common wood-fired and energy-efficient heating choices, and the situations where they may be most useful.

Alternative heating options are discussed in more detail on the energy forums at www.permies.com, www.builditsolar.com, and www.chimneysweeponline.com.

Heating situation: What kind of heat do we need?

Heating needs

Heating methods	Mobile: (camper, tent, boat)	Semi-portable: (Mfr home, yurt)	Rental (apartment, cottage)	Occasional (chapel, ski cabin)	Owned & occupied house	Heat storage / steady	Quick-response	Cooking & preserving	Hot water
Insulation & wind-sealing; curtains	☺	☺	?	☺	☺	☺	☺		
Passive Solar home improvement		☺	?	☺	☺	☺	?	?	?
(Site selection for solar & drainage)	☺	☺	?	☺	☺	☺			
Portable warmers: rocks, beanbags	☺	☺	☺	☺	☺	☺	☺		?
Crowdsource heat: housemates, pets	☺	☺	☺	☺	☺	☺	☺	☺	
Rumford / radiant fireplace	☺	☺	?	☺	☺		☺	☺	
Radiant fireback retrofit			?	☺	☺		☺		
High efficiency woodstove		☺	?	☺	☺	?	☺	☺	?
Radiant space heaters*		☺	?	☺	☺	?	☺	☺	
European masonry heaters**			?		☺	☺		☺	?
Rocket mass heater (RMH)		☺	?		☺	☺	☺	?	?
Floor stoves: Hypocaust, Kang, Ondol	?	?	?	?	☺	☺			?
Radiant floor heating		☺	?		☺	☺	☺		☺
Furnace (forced air)	☺	☺	☺	☺	☺	?	☺		
Boiler (heated fluid or steam)			?	?	☺	☺	☺		☺
Solar hot water	☺	☺	?	☺	☺	☺	☺		☺
On-demand (gas) hot water	?	☺	?	☺	☺		☺		☺
Tea kettles, stockpots, etc.	☺	☺	☺	☺	☺	?	?	☺	☺
Insulated hot water storage	☺	☺	☺	☺	☺	☺	?		☺
Wood cookstove / cooking insert	?	☺	?	☺	☺		☺	☺	☺
Rocket cookstoves	☺	☺		☺	☺		☺	☺	☺
Earthen ovens		☺		☺	☺	☺	?	☺	
Kitchen range / oven (gas/electric)	☺	☺	☺	☺	☺		?	☺	?
Barbecue / grill / smoker / egg-oven	☺	☺	☺	☺	☺			☺	
Sand pan hearth				?	?		?	☺	?
Open fire (air fed, smoke hole)	☺				?			☺	☺

* space heaters: woodstoves, gas logs, electric heaters, fireplaces
** masonry heaters: e.g. Russian fireplace, kachelofen, contraflow, steinofen

Portable Structures and Off-Ground Flats

A mass heater is, well, massive. Masonry heaters are best supported on noncombustible foundations directly on the ground, in the occupied areas of the home.

If your home does not have any ground-floor occupied areas, or is a flat without permission to alter the rooms above, yours is a difficult situation for a mass heater.

Manufactured homes and portable dwellings have additional constraints: they sacrifice energy efficiency and size for convenience, and may have special code requirements for appliances.

It may be easier to build an addition (like a sunroom) onto a manufactured home, or as part of a pull-in shelter for a camper. This makes it easy to support the heater on noncombustible pad or blocks, rather than trying to support a heavy mass on a portable floor system.

We have considered some camper and boat mass-heater designs. In general, most live-aboards and campers are compact enough that insulation (like a room-sized sleeping bag) is an easier path to comfort, and far less weight to transport.

The design particulars for all of these situations are outside the scope of this book, except for Appendix 3.

Domestic Hot Water and Cooking

We enjoy cooked food and hot water year-round, while home heating is a seasonal need. Basic survival cooking can be done on top of the heater's metal bell, or hot water warmed in batches with an open tank or pot. However, providing pressurized domestic hot water is a complex design problem with potentially lethal risks. Most households should have a separate device for summer cooking and for domestic hot water. Boilers and hot water design are beyond the scope of this book, except as discussed in Appendix 3.

Chapter 2

General Design Considerations

Site-specific Design

ROCKET MASS HEATERS are almost always customized to suit a particular building and its occupants. As you examine a building's overall heating needs, orientation, and patterns of use, you may find that the place for a thermal mass heater naturally reveals itself. Sometimes there is only one place for such a large bench to be truly useful. In other cases, you may have to decide between several possible locations based on existing footings and chimneys, clearances, heating advantages, seating arrangements, pathways, etc.

An ideal location for a rocket mass heater is:

- central, where warmth is needed near the home's core use areas (seating/sleeping/dining areas) — not necessarily the exact center of the house
- within sight of daily activities (making the heater convenient to operate)
- on a suitable footing (noncombustible, stout) with access to an appropriate exhaust chimney
- selected with cost in mind: near an existing chimney, footing, thermal mass, planned addition, etc.
- safely clear of combustibles and walkways
- able to deliver heat naturally — by contact (seating, floors), radiant heat (central, line-of-sight), or rising air convection (stairwells, fans, lower areas of split-level buildings).

In real life, we seldom get an absolutely perfect placement. If you are able to satisfy more than half these goals with your proposed location, you're doing pretty well.

Don't forget to consider:

- existing heat sources (kitchen, appliances, sun)
- cold influences (winter winds, exposed walls and windows)
- areas that may benefit from staying cool (pantries, mudrooms, storage)
- areas where people need the most warmth (evenings/offices/nursery)
- passive solar design orientation for free warmth and coolness in season

Heat Transfer

The more effectively a home uses natural heat transfer, the less energy it takes to maintain comfort.

Convection currents in a heated home.

Convection

Hot air rises.

Air is not great at heat transfer; it conducts poorly and causes cooling by evaporation. But if you are generating extra hot air, expect it to rise upward. Controlled vents or stairwells with doors can be a very cheap way to move surplus warmth around a building. (Without a door or vent closure for control, it's easy to overheat the upper floors and start a vicious cycle where nobody is comfortable.)

Air tends to stratify by temperature. Nobody likes a hot head and cold feet, and hot air at the ceiling is basically wasted energy. Warm radiant floors or low benches can help counteract stratification, giving better comfort at lower energy cost.

Warm homes act like a chimney. Air comes in at the bottom and out at the top. This air movement helps ventilate the building, but it can also create problems like cold air currents at the floor, or negative pressure that inhibits proper chimney draft. Homes can also generate negative pressure due to vent fans, winds, and other factors (see "Air Supply," below).

A little ventilation is a good thing, but if you have too much air flow for comfort, give your house a warm cap at the top: weather-seal upstairs windows, ceilings, and attic hatches to stop the leaks, and insulate between ceiling joists. Rising air does carry some heat out of a building, but not much compared with the heat stored in mass and structures.

Radiation

Radiant heat is heat that moves like light. Radiation travels in straight lines, and can be absorbed, reflected, shaded, bent, or re-radiated as it hits things. Radiant heat goes right through air and space, delivering heat mostly to solid surfaces. Obstacles can block a surprising amount of radiant heat. Fire screens, glass fire doors, and heat shields can block 50–80% of the heat from a fire.

Radiant heat works best in line-of-sight. Objects closer to the heat get way hotter (Intensity = $1/d^2$). Move twice as far from the fire, and you get one quarter the heat.

If possible, locate the metal radiant barrel near the center of the heated space, so that all areas to be heated are relatively close.

Radiant heat spreads out in all directions; you may notice round or flat emitters and dish-shaped collectors.

Convection-powered things often have vertical, tube-like, or bell-like shapes.

This line-of-sight principle also makes it easier to monitor the fire from your most-used rooms.

Conduction

Conduction is heat transfer through touch, contact, and within objects (such as heat traveling up a hot spoon). Conduction is almost infinitely more efficient than other forms of heat, so take care not to get burned!

The best way to take advantage of conduction with a rocket mass heater is to sit, lie, or lean on the warm mass. If you place a heater "out of the way" against a wall or level with the floor, be prepared to walk around everyone who ends up lying or leaning against the warm surface.

Conduction is heat transferred in direct contact; look for solids or contoured shapes

Heater Placement

An RMH will redefine your most comfortable hangout areas. Plan on replacing some big furniture, such as a well-used sofa or love seat, unless you're expanding the building. To get the most use out of the heater, consider how you can best make the built-in heated bench into comfortable seating. Don't try to hide the heater, or squeeze it in somehow between old furnishings. This is a massive object. It will dominate the room. Make it a feature.

Rocket mass heaters are most effective in regularly-occupied spaces, where it's convenient to tend the fire and you can fully appreciate the stored warmth.

A heater hidden in a basement or "mechanical room" will never give as much fuel savings as one that is conveniently located near users. If it is not regularly used, it can become so cold that it will be difficult to start in emergencies. And if the heater is in a room that would otherwise be unheated, you've just added a bunch more walls to the workload.

Zone Efficiency

Building codes imply that all rooms need to be heated equally. In practice, most people prefer some rooms to be warmer and others cooler — and storage can definitely be cooler.

Imagine the core of your home as an egg, with the heater as the yolk. At the narrow

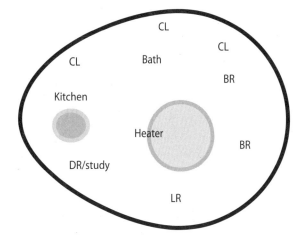

Conceptual home layout around a warm core.

end, kitchen appliances produce their own heat. At the wide end, core living areas cluster around the heater. To protect from excess heat loss, limit exposed exterior walls. A buffer zone of unheated pantries, storage closets, screened porches, and landscaping features serve as a nest, to block cold winds yet admit winter sun.

Radiant Heat

Radiant heat works best in line of sight, so center the heater where it can be seen from most rooms. *The Book of Masonry Stoves* (Lyle) shows several ways to build an oblong masonry heater as a dividing wall, to heat multiple rooms by direct radiant heat.

Sunlight delivers free radiant heat all year. Consider a passive solar design that places the heater mass where it can collect free sun energy in winter (while remaining cool in summer). See Chapter 4 for more details on energy-wise home design.

Heat People, Not Space

When and where do people want the most heat? Sitting, resting, digesting, delicate tasks, sick or small people often need more heat.

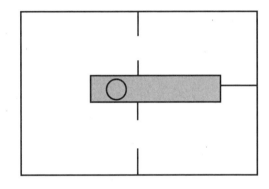

A masonry heater as a partition wall can heat multiple rooms.

Heating priorities: Rooms from warm to cool

Room Type	Description	Core?
Incubator	Nursery, sickroom, sauna, stove room. A warm, quiet room with separate temperature controls offers lifesaving comfort and pleasure.	Varies
Sitting room(s)	Living room, office, TV room, den. Main room(s) in daily use, where occupants hang out, relax, and/or do sedentary work.	Core
Bathroom	Indoor bathrooms and toilets can be heated on demand; heat may be needed to protect plumbing from frost, and to reduce condensation.	Varies
Bedrooms	Small and single people may prefer more heat. Shared beds are warmer, and couples may prefer more privacy.	Core
Occasional use	Spare rooms, formal dining, parlors, guest rooms, hobby rooms, library. Heat on demand, use as buffers when not in use.	Extra
Kitchen	Kitchens create their own heat, and pantries are best kept cool.	Core
Sun room	Greenhouse, solarium, sleeping porch. A suncatching room gathers its own heat during the day, but cools at night.	Extra
Activity rooms	Workshop, exercise rooms, party/rumpus rooms. Activity generates heat. On-demand heat can be used if needed.	Extra
Storage/utility	Closets, pantries, cellars. Cool, dry conditions preserve most items and deter pests. Appliances may generate enough heat to prevent freezing.	Varies
Buffer zones	Mudrooms, enclosed porches, attics, entry rooms, attached sheds. Unheated spaces along exterior walls help insulate the heated core.	Extra

People doing vigorous exercise or sleeping may enjoy a cooler room, and stored items last longer in cool, dark places. We try to heat the living room(s) the most, bedrooms and baths second, and the pantry last. But there's a lot of individual variation.

Consider what shape of heater your household will actually use the most — a barrier between kitchen and living room? A big open seating area? A private nook or heated office desk?

Mass heaters are not a furnace or boiler to heat dozens of rooms, but you can build in some adjustments. The simplest zone control is a room door that can be opened in line-of-sight of the heater, closed when heat is not needed. The simplest portable space heater is a hot rock, brick, beanbag, or potato. Warm-air vents, fans, cushions, and removable heat shielding can also be used to regulate heat delivery.

Hot upper rooms of a house are best used by people who love being warm, or who will close the room door before opening windows.

Building Considerations

Freeze Protection

Plumbing is one major exception to the idea that houses don't need heat, people do. A more complete list might include people, plumbing, batteries, glass canned goods, delicate plants, baby animals, and electronics. Such objects don't have to be kept at room temperature, but they do need to stay above freezing. A heater and "wet wall" in close proximity, in the protected center of the house, can save a lot of grief. A storage cellar might be a great place for cool-storage of canned goods and batteries, but if you don't have one, think about how to locate these items in relation to the heater. It could make the difference in your being able to take winter trips without a house-sitter.

Chimney Design

Using an existing chimney can save hundreds of dollars and days of labor on a heater project. Existing chimneys may need to be inspected and possibly re-lined to match the new heater — an oversized or very rough chimney won't work well.

Chimneys operate on the physical principle of convection. The hotter and taller they are, the more pronounced the draft. If the building is tall, the heater's chimney must be taller, and it pays to keep most of the chimney inside the house where it stays warm. The most effective chimneys, and the easiest to leak-proof, are located near the highest point of the roof.

Some people get excited about the idea of bringing the chimney out the wall instead of the roof, thinking it will make things easier or cheaper. After years of experience with both kinds, we disagree. Issues with structural support, cold chimneys stalling or backdrafting, and leak prevention are all made easier if you exit as close to the roof ridge as possible.

See Chapter 6 for minimum requirements under code, and Chapter 4 for installation advice.

Foundations

Using an existing foundation can save upward of $1000 for a heater project. Many of our example heaters (in Chapter 3) are located on an existing slab, even where it wasn't in the exact center of the home.

If your main room has a suspended floor, it may be easier to locate the heater in a little bump-out addition (like a bay window onto the living room, or a sunroom) where you can easily pour a new footing. Be sure to insulate this addition extremely well, as your heater will now have three exterior walls and may lose a lot of heat.

Example Loads and Footings

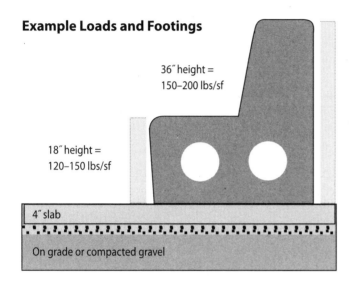

For a more central installation, where the existing floor is suspended wood over a crawl space or basement, the standard practice is to cut away the floor and build up with noncombustible masonry footings from the slab or ground below. In an existing building with an intact concrete basement floor, the footings can be built on top of that (sometimes with a reinforced spreader slab). In areas with frost heave, the accepted best practice is for both house and heater footings to be sunk below frost depth, and stacked up from there. The 2" air gap between masonry and wood framing, usually filled with noncombustible insulation for firestopping, will allow a little bit of wiggle room if the house and masonry decide to tilt different directions in a frost or clay heave.

Building a new foundation below the floor gives you the option of dropping the heater lower in the space, for example lowering the bench to make a radiant-heated tile floor. These sunken heater options will take some attention to clearances and firestopping, but can offer flexible floor layouts.

Weight Loads

If there's a suspended wood floor or full basement under the main room, the heater can be installed in a less ideal heating location (such as an addition, attached garage, or even a daylight basement) for easier foundation work. It's also possible to cut and reframe a hole in the floor, and build up the heater on noncombustible footings from the ground below.

The heater's mass is considerable (3–5 tons in most cases, or about 150 pounds per cubic foot), so it is best supported on solid ground or noncombustible masonry

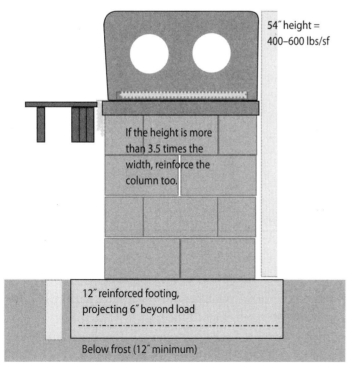

footings. A 4" concrete slab can support up to about 30" of earthen masonry, plenty for most bench-style heaters. Taller projects need thicker slabs and may need reinforcement — see Chapter 6 for more information on foundations.

Size and Use

Smaller buildings and multi-family dwellings are easier to heat.

Large, seldom-used spaces often benefit more from insulation and passive solar design than from thermal mass heating (stored heat is wasted after you leave).

The big deciding factor in using a mass storage heater, compared with other types, is whether the space needs constant heat or just occasional heat. Thermal mass takes a long time to warm up as well as to cool down. If you are not using the space for more than a few hours or days, the delay will be inconvenient, and much of the stored heat will be wasted.

An office where you sit and do delicate tasks like typing for 6 to 8 hours a day might

Home Energy Checklist

Size/Occupancy:
- Core 200-300 sf/person
- Close or rent spare rooms
- Limit exterior wall exposure (compact design; layers; air-locks; storage as insulation; curtains, canopies, fluffy art)
- Controls: Adjust drapes, doors, registers/vents, area heaters, clothing, thermostats, auto-timers, activity levels

Insulation:
- Ceilings
- Walls
- Windows (drapes/layers)
- Floors + crawl space
- Perimeter (foundations, slabs, skirting)
- Clothing and bedding

Weather seals:
Minimum air exchange at least ⅓ home volume/hour.
- Ceilings, attic hatches
- Windows and doors
- Exhaust fans, outlets, light wells, gaps and holes

Appliances:
- Maintain or upgrade: furnaces, water heaters, dryers, cookers
- Radiant and contact heat (solar, heating pads, radiant floors, heat lamps); limit air-heaters and fans
- Keep fuel wood dry

Solar Orientation:
- Windows to catch winter sun, exclude summer sun
- Insulation and/or unheated storage on shady/windy sides
- Thermal mass (water, oil, tile, masonry) to store heat
- Solar heat exchangers for heating, cooling, ventilation

Heating Loads

Most owners want a heater to:
- stay cozy on the coldest day of the year
- keep the pipes from freezing (over a holiday weekend, if possible)
- reduce or eliminate their heating bills
- allow them to survive a power outage or other emergency

benefit from thermal mass heat. A chapel or studio that is only used a few hours each week might not.

My personal minimum threshold for using a mass heater is about 4 hours per day of occupying the space — enough time to operate the heater conveniently while doing other things, especially if there's some residual benefit from stored heat such as keeping electronics dry or plumbing unfrozen.

At more than 8 hours per day, especially when it includes sleeping, I see thermal mass heat as an obvious choice. If the space needs warmth more than 12 hours per day, any other solution begins to look wasteful. Greenhouses, incubators, and starting sheds may need frost protection 24/7 in some weather conditions; so they can be useful places to prototype your first rocket mass heater.

But we don't want to overpay for an over-sized heater that drives people out of the room.

Calculating the heating load of a house is not quite as simple as knowing its square footage. But it may be simple enough to figure out, using your old heating bills and/or a tape measure.

Your old heating bill should show therms (100,000 BTUs) or kWh. Your rocket mass heater could burn up to 12 lbs of wood per therm, or 4.2 lbs per 10 kilowatt hours. Our 8″ rocket heater consumes roughly 20 to 40 lbs per day to heat our 800 sf home.

If you've been heating with a woodstove, a rocket mass heater might use between one half and one tenth the fuel (one quarter is pretty common, but we recommend stocking at least half your previous annual supply the first year until you can see the difference for yourself).

You can also calculate the theoretical heat loss with an online calculator such as the ones at www.builditsolar.org, or www.engineeringtoolbox.com, or work it out by hand. Detailed instructions can be found in Appendix 4. The basic equation to use follows, using the area of each exposed exterior surface, like walls and windows, and the corresponding R-value or (insulation value) of that area.

$$[(Area1/R1) + (Area2/R2) \ldots]$$
$$\times \text{(indoor temp.} - \text{outdoor temp.)}$$
$$\times \text{(time in hours)} = \text{BTUs needed}$$

While you're at it, plug in a few other options (like insulation upgrades, or lower indoor temperatures), and see how they might affect your heating load.

How many BTUs will you get from a mass heater?

Your mass will store heat from any available source: ambient heat, cooking, sunshine, or burning fuel. So you may get some free BTUs any time the day is warm enough, or when the heater is placed well for seasonal solar gain using passive solar design.

The heat available from wood is generally between 6200 and 8700 BTU per pound. (Source: Chimney Sweep's Library, www.chimneysweeponline.com/howoodbtu.htm)

An 8″ heater can burn 20 to 40 lbs of wood per evening, twice that if you can run it all day, for a total of 200,000 to 600,000 BTUs per day. 10,000 to 50,000 BTUs per hour may not sound like a lot, but it's surprisingly effective when combined with the direct people-heating aspect and the correspondingly cooler air temperatures in the home. Plus, you get a bit more per hour in those evening hours when you're there to enjoy it, which is nice.

If you're replacing a large furnace, an RMH likely won't be a stand-alone solution. But it can deliver personal comfort,

Energy Calculations

When calculating the costs of any appliance, there are two phases to consider: the initial cost to buy and install the appliance; and the ongoing cost of running it over time. Ongoing costs include power, fuel, operation, and maintenance.

Payback time is a useful comparison between your current situation and a new alternative. You must know the approximate cost of running both old and new alternatives. Don't forget to factor in your own labor: days off work, or time spent procuring and/or splitting firewood.

Payback Time = Initial Cost ÷ Ongoing Savings

Example: If I pay $1600 per year to run my oil furnace, and I expect to pay $400 per year for cord wood with a new heater, then my ongoing savings could be $1200 per year. A heater that costs $3600 for parts and labor could pay me back in 3 years, and keep saving me money after that.

However, if I only run the heater half the time and keep using the furnace for the other half, I might pay $200 per year for firewood and $600 to $800 to run the furnace (the thermal storage of the bench may still reduce furnace bills slightly). I would only be saving $800 or $600 per year, making it a 4.5- or 6-year payback time.

A heater that I never use, of course, will never pay me back.

Embodied Energy

The embodied energy of any object is the energy used to produce, transport, and prepare it for use. An appliance that is rarely used still costs energy to manufacture, and takes up space. "Green" or "efficient" items do not magically exude worth while sitting unused in the basement.

The embodied energy of a mass heater depends on the situation. Site-sourced earthen masonry, fieldstone, or rubble from nearby demolition can avoid the energy cost of manufacturing and transporting heavy materials. Re-using existing local resources may take less energy than buying new.

Embodied energy can accumulate, however. A portable dwelling (such as an RV or live-aboard boat) uses energy to haul itself around. A mass that can be emptied out for transport (like water tanks) can save energy compared with permanent weight. Low-mass and multi-function systems (like insulation and waste heat exchangers) make more sense in portable dwellings than a solid mass heater.

Energy Independence

In response to cheap fossil-fuel energy, homes have become larger, families broken into smaller units, and local economies have largely been replaced by global commerce. "Independence" has come to mean living alone in a luxurious nest lined with cheaply manufactured goods from far away.

Fossil fuels are a finite resource, however, with increasing costs of extraction and use. Are we ready and able to survive without them? True energy independence means being able to stay warm if the power goes out, and being able to keep things running with local materials and parts.

off-grid security, and substantially offset heating bills, with very little waste. You may be surprised at how much less fuel energy you need, once it's carefully targeted to heating your most important assets.

Older Practices

The original rocket mass heater prototypes were built on dry-stacked rubble footings over well-drained, compacted gravel; some owners continue to use this method for low-cost projects in milder climates. It's important to provide excellent drainage around and under the gravel, and keep damp ground and frost away from earthen masonry.

A few owners have built directly over wood floors, braced up with extra posts and piers to handle the weight. (Reasons not to build directly over wood floors include flammability, code compliance, movement, and longevity. Masonry may trap moisture and rot the wood. Sagging or shifting wood could crack the masonry.) We prefer noncombustible foundations for safety and longevity.

Put some care into foundations: they're hard to fix later.

Air Supply

Short answer: No, we don't favor outside air for combustion. It creates more problems than it solves.

Fresh air is good for you and good for your house. But the fire needs warm air as much or more than you do; cold outside air can reduce combustion efficiency, especially if it also makes it harder to properly control the fire, load wood, or shut down the draft afterwards.

Let the house breathe; fix any negative pressure problems (they're bad for other reasons too); and consider preheated make-up air if you need it.

Stale air is a very poor way to store heat, especially considering you need to ventilate the house at least one third of its volume per hour anyway. Radiant and conductive heaters offer better comfort at lower air temperatures. Lower indoor air temperatures means less heat loss from both conduction and ventilation.

If the home is so badly designed that building an outside air supply onto a downdraft heater sounds simple by comparison, see Chapter 6, for warnings and recommendations.

The Seat Test

The final test for any heater design is the seat test, or as Ernie usually calls it, the "butt test." How comfortable is it to sit on the heater? Try it. Test the proposed height, the arm-rests and back slope, the proposed cushions. Get your sweetie to sit on it. If your household won't sit on the heater, much efficiency is lost.

Get your favorite interior designer involved. What social and ergonomic factors could encourage people to sit on the heater? Cushions and aesthetics are important to many people. Consider "conversation groupings" where the heater faces another sofa or comfortable chair. You could place work desks with seat or feet near the heater or make benches long enough to offer a full-body heating pad or massage table after a long day.

To sum up, a well-designed radiant heater is:

- appropriate to the household's heating needs and patterns of use

- supported on suitable foundations, with safe clearances to combustibles and a good chimney
- located centrally in the occupied space, with line-of-sight to the most-used rooms
- comfortable and attractive, encouraging direct contact (seating, floors, etc.)
- cost-effective, offering the best long-term return for the least up-front cost, for example by taking advantage of existing chimneys or footings when there is no conflict with overall performance

A 4-ton heater is hard to hide. Make it a feature.

Materials Considerations

The original rocket mass heaters were made with clay brick and earthen masonry. Traditional masonry heaters used clay materials for their plasticity, ease of maintenance and repair, excellent heat tolerance, and thermal storage performance. Modern materials may be more expensive, harder to repair or rebuild, and have more embodied energy.

A warning: Portland cement is not suitable for high-temperature fireboxes.

All lime-based products, including Portland cement, degrade to crumbling powder in the heat of a clean fire: 1200°F to 2000°F (600°C to 1100°C). Concrete can be used in footings and heat-exchange areas, with some caveats (cement stucco is not compatible over an earthen core; concrete is less comfortable than earthen masonry for seating). See Appendix 1 for more details.

Refractory cements are a different option; they may be used where required by regulators. Based on work to date, we suggest that dense refractories be rated for at least 2100°F/1150°C; light/insulating refractories used without dense lining be rated for at least 2500°F/1350°C.

Modern refractory cements are harder, stronger, more brittle, and harder to repair than clay. Strong mortars can break bricks under stress, complicating repairs. Each type of cement is unique, and responds differently to water, time, heat, and other factors. Each type requires specific curing conditions. Read the material data carefully, practice, and heat-test your results before installation.

Expansion Joints

Materials expand with heat. If the hot firebox is trapped on all sides by the cooler casing, something will crack. We need a flexible material or gap for thermal expansion, called an expansion joint.

With the original design, we simply heat everything up while the clay-based core and casing are still pliable to create the necessary expansion allowance. If there are any cracks, they can be filled while hot, and thus stabilized, before doing the finish work.

Where it may not be possible to fire the heater while the casing is still pliable, build in expansion joints. Use flexible, high-temperature insulation such as ceramic-fiber refractory blanket, rock wool, braided fiberglass gasket, or even stabilized perlite.

Provide a ⅛" or larger expansion joint around the firebox and any square, high-temperature masonry (such as the manifold) and at any metal corners or edges (such as access doors).

Masonry insulation dust is insidiously dangerous due to its fine structure. Keep masonry materials damp while working, or wear a good-quality respirator.

Lifestyle Considerations

At-home Time: A manually operated heater is reasonably convenient for people who actually live in their homes. Active parents, the self-employed, retired people, or those who enjoy quiet evenings at home can run a heating fire without interrupting their lives. A rocket mass heater should be located in a room where people like to spend time. If you don't spend much time at home (fewer than 3 evenings per week), a wood-fired heater may not suit your lifestyle.

Emergency Heat: Owners sometimes invest in a wood heater just for supplemental and emergency heat. Firing up a wood heater a few times per week can save substantially on furnace bills, and keep you cozy during power failures. Unused mass heaters in an unheated basement, however, are slow to warm up, harder to start, and therefore less reliable in emergencies. Will you maintain a supply of dry firewood "just in case?" Dry fuel is critical for both efficiency and safety. For a true emergency heater, consider an easier-to-start, quick-delivery radiant heater, such as a certified wood-burning stove, and make sure this heater is located close to the core areas needing heat in an emergency (plumbing wet wall, or wherever the household will gather to weather an emergency).

Occasional Use: Guest rooms, chapels, and vacation cabins may need heat only occasionally. In a place that is only used for a few hours per week, heat stored in mass is largely wasted. The first priority for occasional-use spaces is excellent insulation and passive solar design, which benefit all buildings, regardless of fluctuating use.

Heaters for briefly occupied areas can be more responsive and less efficient, since they are used less. Responsive heaters have thinner walls or smaller mass, and are configured for easy start-up and quick heat delivery rather than heat storage.

Automation: If an area or object needs to be kept at temperature for long periods without an operator (for example an irrigated plant nursery, institutional housing, or plumbing), consider an automated heating system such as a furnace, boiler, or electric heat.

Community Operation: A heater that serves more people requires proportionally less investment from each. Three or four adults sharing one home have more hands for fuel processing and maintenance, and can take turns tending the heater with fewer missed evenings. Rocket mass heaters can be an important part of heating multi-family dwellings up to 2 stories tall. To heat larger spaces, consider a high-capacity masonry heater, or zone heaters for each family.

Multi-function: An efficient heater may not be an efficient or comfortable all-purpose device. People with very small or efficient buildings may want a heater that serves as many purposes as possible: cooking, baking, heating, and perhaps hot water. It's important to note that these heating needs occur at different times (for example people want hot water and cooked food in summer, when household heating is undesirable). Options to divert unwanted heat, such as a heat-exchange bypass valve, and a heat shield around the barrel with venting to outdoors, may be useful. A better option for serious cooks may be to create a separate kitchen or outdoor summer kitchen.

Values and Goals: Even the closest people have different priorities. For some owners, a single goal or "deal-breaker" defines success or failure. Families may disagree about the relative importance of visual appeal, resale value, up-front cost, ease of operation and maintenance, and energy efficiency. The heaters presented here meet several possible goals, including quick response heat, efficient long-term heat, adjustable fueling, nontoxic materials, and durable built-in furniture.

Some elements of the heater can be changed without affecting performance (colors, surface textures, decorative metal bells). Other changes do affect performance (hiding the heater, poor chimneys, height changes, extra functions like hot water or cooking that operate on a different schedule than winter heating). We recommend that the household members agree on the top two or three priorities, list others as optional, and choose a proven design that addresses the top priorities.

Chapter 3

Design Examples

THE FOLLOWING EXAMPLES show how particular owners have designed and finished their heaters. We have included rough floor plans so you can see how each heater was placed in relation to the existing building. These examples are in the northern hemisphere, so the sun is from the south.

Notice how the heaters tend to be located near the core of the house: in living areas, larger rooms, or against an interior wall between living and bed rooms. Look for multi-functionality — heaters that are also seats, beds, or well-placed for convection heating, cooking, and warming food.

Some heaters were placed on an existing slab footing, or in a new addition for easier foundation work, rather than centrally to the house. The key provided here shows the shaded colors that indicate the pre-existing floor type for the occupied areas being heated.

| Concrete slab floor |
| Suspended wood floor |
| Other/no floor (earth, gravel) |
| Existing chimney/fireplace |

Annex 6" Heater photo © 2009 by Kacy Ritter.

Annex 6" Rocket Mass Heater

Heater Summary

HEAT-EXCHANGE MASS:
Duct size: 6" ID
Duct length: 25 feet (excluding 20-foot chimney)
Thermal storage material/size/weight: earthen masonry bench, 30" x 7' and 27" x 7', 6000 lbs

COMBUSTION UNIT:
Heat riser height: 47"
Fuel feed height: 15"
Burn tunnel length: 21"
Firebox opening size: 5.5" x 6"
Manifold: square brick plinth with metal pipe fitting, cob and earthen plaster

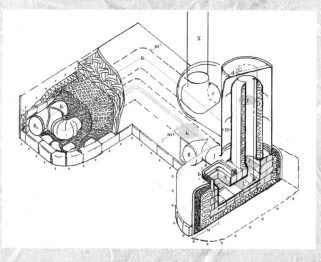

Site Details
Project dates: 2008–2009
Location: Portland, Oregon
Building size: 900 sf
Chimney height: 27 ft
Foundations: existing 4" concrete slab

Annex 6″:
Compact Hospitality Bench, Case Study for Permits

Building Considerations

The building is an attached rental apartment (an "annex" to the main house), circa 1951. The climate is maritime Western Oregon: an 8- to 9-month rainy season, occasional freezing, near-constant cloud cover, with peak cold in the months of January and February.

Total square feet to be heated are about 900 square feet (sf). The layout is not ideal for heating: lots of exterior wall surface area, no sun exposure, many single-pane windows, constant damp conditions, poor wall insulation. Even after restoring the storm windows and adding some insulation, the heat loss from this many-walled 900 sf apartment might be similar to that of a square building of 1500 sf.

The existing furnace was tucked away in an adjacent space, with warm air delivered through attic ducts. Gas bills were typically $100 to $150 per month, supplemented with electric area heaters due to poor heat transfer. The warm ceiling and chilly slab floor combined for discomfort winter and summer. Mold problems were evident along windowsills, exterior walls, and in the bathroom.

In the main rooms, a 4″ existing slab floor (grey) served as foundation for the heater. (Even though the heater spans two separately poured sections of slab, no shifting or cracking were observed in either slabs or heater over the 4 years following installation.)

The attached bedroom with its suspended wood floor is raised up by an 8″ step, creating a slight warm air transfer to the bedroom.

Design Considerations

The heater was located in the largest available space, the living room, with line-of-sight to the dining area and a bedroom. The L-shaped bench design echoes Moroccan hospitality divans, serving as both seating and guest sleeping space. Each leg of the L is long enough for a full-sized adult to sleep comfortably. A sturdy matched bench/

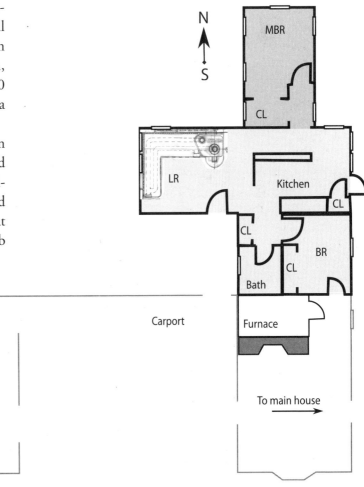

coffee table can be pulled up and covered with cushions to create a double bed for guests.

We chose a 6" heater partly due to the small space to be heated, partly for space concerns, with some thought to cost of parts, and partly because the property included many well-established orchard and forest trees that produce copious branch wood for pruning. A smaller firebox is very convenient when burning small pole-wood. In retrospect, given the exterior wall exposure and established mold problems in some of the farther rooms of the house, a larger heater might have been useful.

Typical burn pattern: The heater typically burns 4 hours each evening, up to 6 hours in extreme cold snaps (teens or below zero lows). An 8" heater might handle the same heating load with only 2–3 hours of burn, or by firing on alternate days.

During the shoulder seasons (spring and fall), the heater might be fired two or three times per week. During the hottest months, the un-heated mass has a cooling effect, soaking up daytime heat for slow release during the cooler nights. After the heater was installed, we discovered we no longer needed the portable fans we used during previous summers.

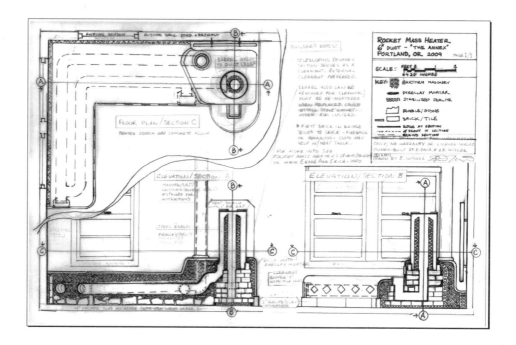

Bonny 8" Convection Bench all photos from 2011-2012 DVD project "How to Build Rocket Mass Heaters with Ernie and Erica," © 2011 Calen Kennett and www.villagevideo.org.

Bonny 8" Rocket Mass Heater

Heater Summary

Heat-exchange Mass:
Duct size: 8" ID
Duct length: 32 feet excluding chimney
Thermal mass: Earthen masonry bench on brick footing, 24" to 50" wide, 14 feet long

Combustion Unit:
Heat riser height: 54"
Fuel feed height: 16"
Burn tunnel length: 26"
Firebox opening size: 7" × 7.5"
Gap above heat riser: 2"
Gap in manifold: 4"
Manifold: octagonal brick plinth with earthen plaster

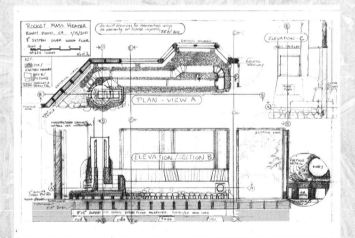

Site Details
Project dates: 2010–2011
Location: Northern California
Building size: 5000 sf
Chimney height: 25 ft
Foundations: reinforced wood floor

Bonny 8": *Air Channels, Multi-story House*

Building Considerations

The building is about 5000 sf. The main rooms have cathedral ceilings about 20 feet tall. Both wings of the house have "turret" staircases to second-story bedrooms and a third-story guest room. The heating season in northern coastal California has little snow, but frequent fog and damp chill.

The existing forced-air heating system ductwork was contaminated with mold and mildew. The occupants had been heating with a fireplace insert (exterior masonry chimney) and a small supplementary woodstove in the far wings, using about eight cords of wood per year. The insert did not release or store much heat, and the house stayed uncomfortably cool most of the winter.

Design Considerations

For once, the heating factors and seating factors happened to agree on the best placement of this heater. The main living room was on the lowest floor in this building. The most logical place for permanent seating (opposite the existing fireplace) was on the same side as the main stairwell doors, offering the opportunity to easily circulate warm air to upstairs rooms. The southern exposure was shaded by evergreen trees, so solar gain was not factored in. (Without the shade trees, this bench would receive more solar warmth in summer, but not in winter, the opposite of desirable solar gain.)

The owners wanted to keep the existing hardwood floors in case they had reason to remove the heater. They also wanted to heat the entire house with a single heater, without using their moldy air ducts.

These two considerations combined to suggest a solution from earlier European masonry heaters: raise the heated masonry up on feet or air channels, to allow room air to pass below the heater. This room air simultaneously helps to cool the wood flooring, while extracting useful heat from the

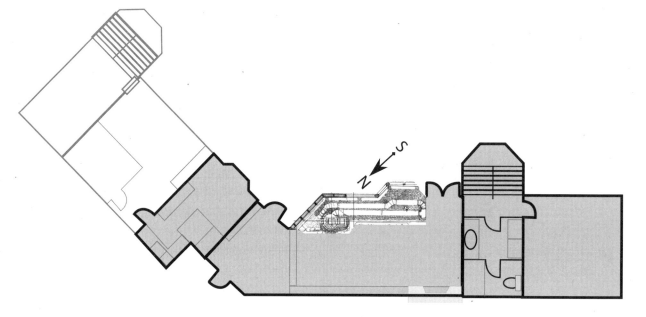

underside and back of the heater, releasing warm air for circulation to distant rooms.

The redwood back is vented at the top to promote this circulation; warm rising air mixes with cool air from the windows, reversing the room's natural convection. The heated air rises along the same wall as the staircase to the upstairs bedrooms, creating easy natural heat flow with the opening of a door.

Of course, the wood floor had to be reinforced to support the added weight of the heater. An 8″ by 8″ beam was added directly underneath the front edge of the new heater, supported by four new piers.

The presence of small children and frequent guests, and aesthetic preferences, dictated the extent of earthen plaster around the barrel and the artwork painted onto it with natural pigments from around the world.

Cleanouts are hidden behind tiles (such as the octopus tile at front right), or behind the barrel area.

The occupants still use the insert and woodstoves occasionally, for supplemental heat or entertainment, but the rocket heater is the primary heat source, operating every day in winter (typically 4 to 6 hours in the evenings, or up to 8 hours when someone is home during unusually cold weather). The new heater serves as the main heat source for the building. The combined firewood usage for this heater plus all other fires has dropped from eight cords to two cords per year.

This heater was documented in the Village Video production, "How to Build Rocket Mass Heaters with Ernie and Erica." Photo Credit (preceding pages): Calen Kennett, www.VillageVideo.org

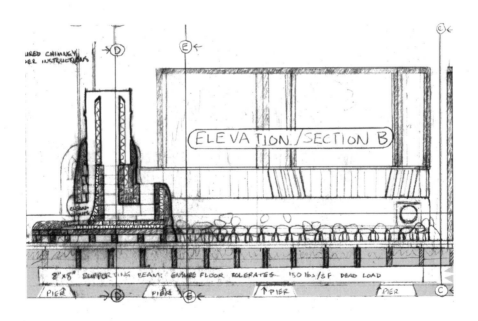

Cabin 8" Rocket Mass Heater

Heater Summary

HEAT-EXCHANGE MASS:

Duct size: 8" ID

Duct length: 20 feet in bench

Thermal mass: fieldstone and rubble with earthen mortars, 30" to 36" wide, 8 to 9 feet long, seat height 19" with raised back 48" to 60" tall

COMBUSTION UNIT:

Heat riser height: 48"

Fuel feed height: 16"

Burn tunnel length: 24"

Firebox opening size: 7" × 7.5"

Gap above heat riser: 2"

Gap in manifold: 3.5"

Manifold: two-barrel style, metal manifold form with field stone casing

Site Details

Project dates: 2011–2012

Location: Okanogan Highlands, Washington

Building size: 800 sf

Chimney height: 15 ft

Foundations: slab on grade

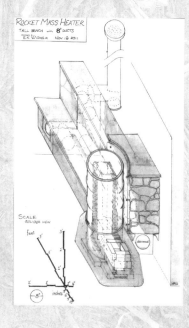

Cabin 8": *Rocket Mass Heater*

Building Considerations

The cabin is a compact 800 sf, located at 3500 feet elevation in the Okanogan Highlands (pine and sagebrush country near the Canada/US border). As in most northern border states in the continental US, the winters are cold, with snow on the ground from October through April many years.

Originally a 24 × 24 foot garage with slab floor, with a later 12 × 24 addition of bedroom and bathroom over a crawl space. All exposed walls, ceiling, and floor have R-30 to R-40 insulation, and the crawl space has skirting and insulation. Most windows face south for better sun advantage.

Design Considerations

The heater is located on an interior wall between two doors: the entry door to the living area, and the bedroom door. (This is not an ideal seating arrangement: anyone using the bench is liable to get bumped by the door swing, so it tends to collect coats and hats and packages more than we'd like.)

The small available space and the presence of a stout stem-wall on this side of the foundation dictated a taller heater with enough volume for three runs of heat-exchange ducting. The stone and tile mantle behind the barrel serves as both heat shielding and additional heat storage.

Local geology provides mainly volcanic silts, fractured bedrock, and glacially deposited granite. Using fieldstone made far more sense than importing earthen materials. Other aesthetic choices include pine trimboards on bench and shelf/back (suitable for hanging stockings). For safety, the shelf above the hot barrel is not wood but Saltillo tiles in a similar size and color.

In winter, the heater is run about 4 to 6 hours per day, usually in the evenings. During the coldest three months of the year, it burns about 20–35 lbs of wood per day. During shoulder seasons, the same amount of wood typically lasts 3–4 days, with the heater operated either for shorter times, or only a few nights each week. We often stop firing by May and just use cross-ventilation.

In summer the heater is not fired. Instead, it "stores" the evening cool and buffers against daytime heat. Temperatures can be 10 to 20 degrees cooler indoors during the peak of a summer day (90s to 100°F outside, high 70s to low 80s inside).

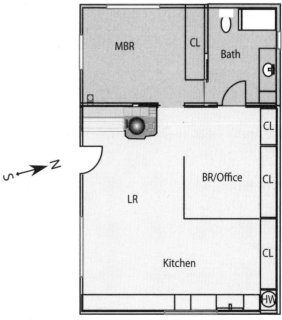

Daybed 6" Rocket Mass Heater

Heater Summary

Heat-exchange Mass:
Duct size: 6" ID
Duct length: 15 feet in bench
Thermal mass: earthen masonry and rubble, 11 feet long including 7' by 4' bed.

Combustion Unit:
Heat riser height: 48"
Fuel feed height: 16"
Burn tunnel length: 24"
Firebox opening size: 6" × 5.5"
Manifold: brick and earthen masonry

Site Details
Project date: 2010
Location: Portland, Oregon
Building size: 120 sf
Chimney height: 7 ft
Foundation: slab on grade

Daybed 6": *Bypass Damper for Intermittent Heating*

Building Considerations

The building is a one room, tiny guest cottage in maritime Oregon, about 11 by 11 feet. The project serves as a testing ground for the owner's interest in natural building. It converted an ordinary stud-framed shed into a cozy retreat with an earthy, natural feel to it; it has straw-clay insulation, earthen plastered interior walls, natural paints, recycled windows, and this compact, heated bed.

Building Considerations

The heater had to fit in available space without impeding the exit, so the barrel is tucked in the corner, and there's a sculpted "berm" to keep the mattress off the barrel. The barrel only gets to about 200°F at the bottom anyway, but safety is especially important in guest sleeping quarters.

Because the "guest shed" may not be heated year-round, the project has a bypass for easier lighting in a "cold start" situation. (See Chapter 5 for cold-start lighting instructions.) The owners sometimes enjoy sleeping in the guest shed themselves, and run the heater about twice a week when the building is occupied.

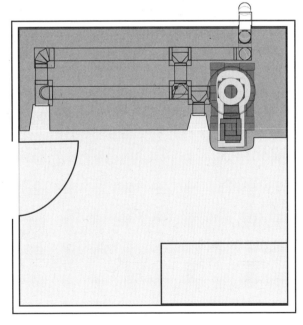

Ellie the Elephant Heater: finished bench photo by project owner.

Mediterranean 8": Warm Climates

Heater Summary

Heat-Exchange Mass:
Duct size: 8" ID
Duct length: 17 feet in bench
Thermal mass: cob with fieldstone, bench 3 feet wide, 8 feet long; seat height 15" with hollow wooden back. (Total length, 11 feet)

Combustion Unit:
Heat riser height: 48"
Fuel feed height: 16"
Burn tunnel length: 24"
Firebox opening size: 7" × 7.5"
Gap above heat riser: 2"
Gap in manifold: 3.5"
Manifold: two-barrel style; polished upper barrel

Site Details
Project date: 2013
Location: Southern California
Building size: About 1200 sf
Chimney height: About 20 feet
Foundations: raised concrete footing in crawl space

Design Examples

51

White heater with blue tile: built by Adiel Shnior, Amit Pompan, and Nitzan Isrovitch in 2013. Photograph © 2013 by Adi Segal, www.adi-segal.com.

Mediterranean 6":
Warm Climates

In warm climates, including arid regions like the Middle East or the American Southwest, heating loads are typically smaller, but may be divided into extreme days and nights. Cold-season heating needs may be intermittent rather than steady, with one or two cold snaps that really call for heat. The occupants may also want heat for personal comfort even when it is warm outdoors; for example as a personal heating pad after exertion, or when coming into a cool house after a hot day outdoors. Chimney exhaust does not draft so well if the outdoor air is also comparatively warm, so mild-climate heaters may need to have extra draft built in for successful cold starts.

In these climates, thermal mass alone will increase comfort. Traditional architecture like adobe takes advantage of the greatest benefit of thermal mass: to stabilize the spikes between daytime highs and nighttime lows, for greater comfort even while using no fuel at all.

For warm climate designs, we prefer a relatively short, linear bench with a double run of pipe. Placing the chimney beside the barrel gives a draft boost during startup, and it's also relatively easy to include a bypass if desired. A bypass not only facilitates starts in warm weather, it also allows comfort heating on cool evenings without charging up the mass, useful when days are too hot and evenings just slightly too cool for comfort.

Building Considerations

Re-framing the floor and adding a noncombustible foundation turned out to be simple compared with the chimney installation. The building's back rooms had been added later, along with a second, shallower roof right on top of the first one, making it difficult to fit the chimney between the complicated mess of rafters anywhere but at the edge. There was just barely room to fit the firebox between the noncombustible fireplace and the exhaust ducting, with extensive heat shielding (the wall surface of faux logs was removed behind the heat shields to give better air movement, which took an extra two days).

Design Considerations

This compact heater layout fits the space nicely, and preserves the main functions of the room and existing fireplace. A local plaster artist did the finish work, adding the whimsical elephant face to the arm rest. The owners and friends now call it "The Elephant Bench."

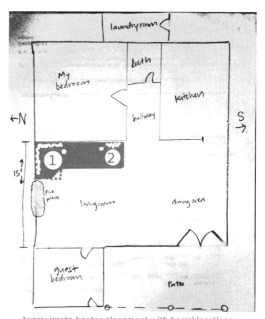

Approximate heater placement with barrel locations:
(1) Owner preference,
(2) theoretically more efficient

Heat shield / high clearance areas

Phoenix 8" Rocket Mass Heater

Heater Summary:
Bench:
Duct size: 8" ID
Duct length: approx. 30 feet in bench
Thermal mass: Concrete block and "brick" exterior, with earthen infill

COMBUSTION UNIT:
Heat riser height: 48"
Fuel feed height: 16"
Burn tunnel length: 24"
Firebox opening size: 7" × 7.5"
Gap above heat riser: 2"

Site Details
Project dates: 2011–2013
Location: Inland Washington
Building size: 1800 sf
Chimney height: 15 ft
Foundations: cinderblock on grade
Manifold: two-barrel style with brick and earth infill

Phoenix 8": *Masonry Bench with Mineral Soil Infill*

Building Considerations

At 3800 feet in the Okanogan Highlands, residents expect about six months of bright snow, and two to three months of dry heat. Local geology is mainly volcanic silt and glacial till, covered in pine and sagebrush (lovely, but not great soils for earthen masonry).

The original structure was a small shed used for camping vacations. Successive additions expanded the building into an 1800 sf year-round home. The woodstove remained in use as supplemental heat, burning up to four cords per year to offset the cost of running a propane-fueled central furnace.

Design Considerations

The furnace already occupies the central place of honor, and the occupants saw no reason to get rid of a perfectly functional furnace. They also kept the woodstove until the new heater proved itself.

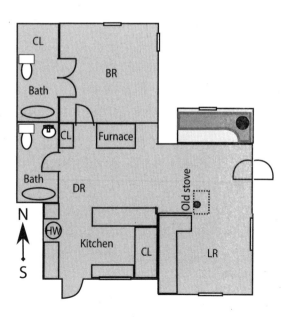

Adding weight to any of the various floors seemed likely to cause uneven settling of the structure. But there's always room for another addition. The owner built a cinderblock footing onto the north side of the living room, and laid out the heater with the barrel on the east side and the chimney exiting through the north wall.

Around the ductwork, he dumped several tractor-scoops of local, silty soil, then laid bricks and plywood across the top to prevent the cats digging in it. After the addition was weathered in, the owner partially removed the adjoining wall to open the heater alcove into the living room.

During the first winter, alterations included insulating the little chimney for better draft, and then replacing it after an ice dam toppled it. Uneven settling of the loose fill, along with the chimney collapse, led to draft problems that suggested the manifold and/or pipes might be leaking.

The following spring, the owner dug out the installation and re-designed it, using brick to create a facade in front and laying the pipes closer together. Brick is also planned to extend well above the barrel on the corner walls, offering some additional heat storage. The exhaust was rerouted out the gable-end wall (near the front door) and a weathercock-style chimney cap added above the ridge line.

The heater is used in the same way as the woodstove: to offset the furnace bill. It has reduced the furnace bill by about 80% (one-fifth of previous). During the colder months, the furnace comes on about 4 am to keep the house at 68°F. The heater uses less than half the wood they were previously feeding their small woodstove to offset the furnace use, and is saving a lot more on utilities.

Greenhouse 8" Rocket Mass Heater

Cob (earthen masonry) softens when exposed to constant damp conditions. This bench was used as a step to water plants hanging overhead.

We replaced the crushed pipe, then used brick and sandstone slabs (next page, right) to create a load-bearing masonry box around the pipe.

We also reworked both heat riser and chimney for better draft. We sealed leaks, enlarged flow volume in the manifold, and improved insulation around the heat riser.

The exit chimney was originally horizontal, and caught a lot of wind gusts from a large building nearby. We rebuilt the chimney as vertical as possible, inside the greenhouse for warmth, then made a dogleg out the end wall and back up, ending above the roof.

Site Details

Project date: 2012
Location: Ohio
Building size: 600 sf
Chimney height: 15 ft
Foundations: asphalt parking lot
Firebox size: 7.25 by 7.25 inches feed, tunnel, and heat riser; 16" tall feed, 48" heat riser
Manifold: sculpted cob with sandstone cap

Greenhouse 8":
Damp-tolerant Detailing
Building Considerations

A greenhouse heater can be a great, low-stress first project. Greenhouses come with extra challenges, however, because they are:

1) damp, which can cause mold, rot, corrosion, and structural problems.
2) poorly insulated and have high solar gain, leading to dramatic temperature swings, venting, and air pressure effects.
3) temporary structures which may not have footings or through-roof chimney options.
4) often unoccupied during heating hours, making manually operated heaters less convenient.

Design Considerations

Damp and impermanent conditions argue against earthen masonry in greenhouses. The box-and-fill method can be used instead to make raised beds. The box may be rot-resistant wood, brick, or other masonry. The ducts can be stainless or ceramic for longer-lasting benches, or cast in place. In any case, the ducts should be protected from garden tools with a layer of tile, stone, etc. Fill can be tamped dirt (subsoil) to add mass around the pipes below the planting levels. Some greenhouse owners bury the heating channels under the floor for more flexible space.

Cold starts are a routine consideration for a greenhouse, due to damp conditions and variable heating schedules. Greenhouses may also have severe negative pressure due to unbalanced venting. See Chapter 5 for cold-start options to boost draft in adverse conditions.

Safety: Don't be tempted to vent exhaust indoors; even the cleanest exhaust is not safe to breathe.

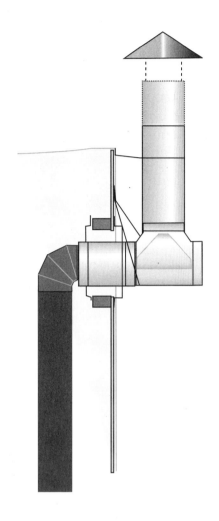

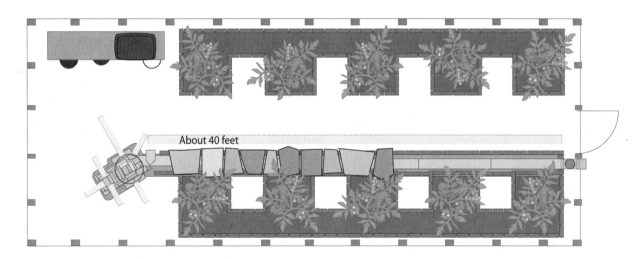

Option 2: Wood or hardiboard planting box

A lower-cost alternative for areas where brick and sandstone are scarce. (180° U-turn pipes give better draft and heat distribution)

Considerations for a raisedbox planting bench:

- Allow safe clearances around firebox and bell
- Heat shield or replace nearby combustibles
- U-turn locates chimney near barrel for draft
- Use 4" masonry spacers between box and bench pipes
- Dirt or gravel fill to 8" above pipes, with a stout divider over pipes (tile/brick/wood)
- Place topsoil or planting trays above divider

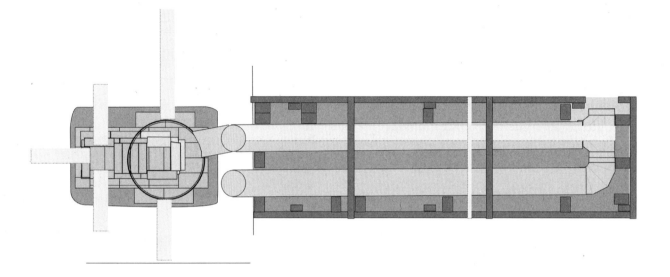

Chapter 4

Step-by-Step Construction Example

The first step in any installation is planning. All the chapters of this book are important for planning a successful rocket mass heater project. If you've skipped ahead to this chapter, please do read the rest of the book before doing anything permanent.

This chapter shows the steps from a finished design to a finished, operating heater. We selected a J-style firebox as our example installation because it's popular and reliable, especially for owner-builders on a budget.

Our example heater is an 8″ system; it uses 8″ ID (inside dimension) stovepipe, and brick channels that have the same cross-sectional area (CSA). The CSA, or flow area, for an 8″ diameter system is about 50 square inches.

The firebox is built of firebrick with refractory blanket insulation. The bell and manifold are 55-gal (200-liter) steel drums, about 23″ diameter. The bench has metal-pipe-lined heat-exchange channels set in monolithic earthen masonry (cob). The exhaust comes out near the bell, into a manufactured chimney about 20 feet tall.

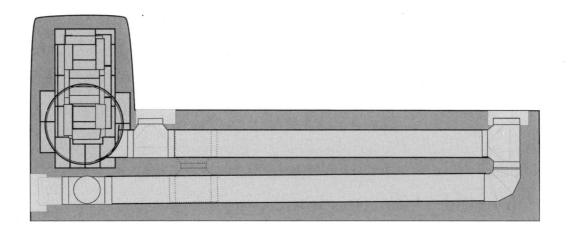

Foundation work is not shown in our illustrations. This type of heater can sit on a 4″ concrete slab on grade, or a similar noncombustible footing suited to local drainage, frost, and seismic considerations.

In the sections that follow, we will tell you what tools and materials you need and what planning and prep work is involved. We will then guide you through the steps of building the combustion unit and the thermal mass heat exchanger. We will tell you about "firing in" and give you some ideas about finishing.

Tools and Materials

Tools to Suit the Team

Mixing and moving heavy masonry materials is undeniably work. People often suggest big tools to automate the cob-mixing process: concrete mixer, rototiller, small tractor. Effective use of this kind of equipment takes experience and planning.

Sometimes more power just means bigger, faster errors. We rarely use heavy equipment with mixed work crews, due to noise and hazard. That said, a cement mixer can work on either powdered dry material or soupy wet batches of mortar. Most cement mixers can't provide shear mixing (smearing/stirring) for tough cob batches. Very wet cob can be pre-mixed ahead, then allowed to dry to a stackable consistency.

Rototillers reportedly do OK; cob mix can be contained in a pit a little deeper than the tiller's tine depth. Bobcats or small tractors can do both shear and bulk mixing, especially with a skilled operator. An inexperienced tractor operator may not be able to keep the batch together without help or work safely around helpers.

Most able-bodied people can tarp-mix cob at a similar rate as novice equipment operators. It's a matter of preference and practice. With due care and good music, many people enjoy the exercise.

The best tools and methods are those that work for your crew.

Building Materials

Unit Masonry:

- Firebrick (9″ by 4.5″ by 2.5″) and half-firebrick (9″ by 4.5″ by 1.25″)
- Rubble (broken concrete, local fieldstone, or brick, for footings and face)

Tool List

Measuring and Marking:
- Tape measures, pencil and pad
- Masking tape/chalk line
- Level (plumb line, optional)
- Squares/angle bevel

Masonry and Mortars:
- Buckets
- Shovel(s)
- Tarp(s)
- Mortar trays
- Paintbrushes (2" to 4" size)
- Water mister/sprayer
- Mason's trowel(s)
  - Concrete float
  - Plaster float(s)
  - Paint scraper or corner trowel
- Hammer: Framing/2# sledge
- Mallet or wooden handle
- Cold chisels/brick-set
- Wheelbarrow/dolly

Metal:
- Tinsnips/heavy cutters
- Crimpers for duct/stovepipe
- Pliers (brake or flat-seamer optional)
- Hacksaw or grinder
- Screwdriver and bits
- Wrench, pliers, gloves

Wood Work (Alterations):
- Saw (circular/flush-cut)
- Framing hammer
- Drill and screwdriver

Cleanup and Safety:
- Safety glasses/goggles
- Gloves (rubber/leather)
- Dust mask(s)/respirator
- Hose/outdoor wash station
- Buckets
- Broom, mop, rags
- Vacuum cleaner (Shop-Vac)

Power Tools/Upgrades:
- Electric drill with paddle mixer for mortar, clay slip
- Wet-dry shop vacuum
- Circular saw, hand-held grinder, tile cutter, and/or table saw with blades for:
  - masonry (diamond grit blade)
  - metal (grinder/cutter)
  - roofing (site-specific)

Experienced operator for all power tools, especially if using larger equipment like a cement mixer, tractor, bobcat, or rototiller.

- Mortars: Clay slip (or thin-set refractory mortar)*

Earthen Masonry:

(See Chapter 2, "Weight Loads" and Appendix 1 for more about earthen masonry and workable alternatives.)

- Clay (recycled pottery clay, raw fireclay, or local clay soils)
- Aggregates: masonry sand or traction sand, plus small gravel (¾"-minus/pit run)
- Straw and fiber for outer casings
- Insulation: Perlite base, DuraBlanket (ceramic fiber/rock wool rated for over 2000°F)

Metal:

- 55 gallon drums: two matched steel drums in good condition, clean of paint and debris
- Matching steel barrel lids and band-clamps (food-grade barrels often come with clamp-on lids)
- 8" ID stovepipe (or ducting) for channels in the following shapes:
  - 1 elbow
  - 3 capped T's for cleanouts (each T needs one matching closure cap — preferably uncrimped stovepipe caps that fit over the pipe, not into it)
  - 2 straight 5' pieces and 3 straight 2' pieces (or approx. 16 feet total straight pieces)
- 8" ID chimney (measured to fit the structure — dimensions will vary from project to project):
  - stovepipe/manufactured chimney long enough to reach from bench to ceiling
  - insulated class A chimney sections from ceiling through roof
  - matching through-roof fittings including screened chimney cap

A Clean Barrel for the Bell

The bell is steel or stainless steel, airtight, and free of unsuitable paints and debris. You can approach finding the right barrel in a number of ways:

1. Get a recycled barrel, such as a food-grade 55-gallon drum. Remove the paint by abrasion, which means sand-blasting, sanding, or grinding; OR burning the barrel outdoors. See sidebar, "Pocket Rocket," later in this chapter.
2. Salvage an old water-heater tank or similar cylinder: remove all fittings and contents. Taking steps to prevent pressure explosion, leave one end sealed and cut the other open. Remove surface paint as above. Interior enamels may be left in place or removed.
3. Purchase a clean, unvarnished stainless steel drum directly from a manufacturer. (If varnished, remove varnish by abrasion or burning, as above.)
4. Commission a custom stainless or weathered-steel cylinder (or a similar column shape, such as an octagon or hexagon) from a qualified local welder. To maintain performance comparable to barrels we've tested, the steel should be at least 12-gauge, 20 to 24-inch inside diameter, 28 to 40" interior height, with airtight construction including an airtight lid or sealed top.

Protect the clean, bare metal with a heat-tolerant coating. You can use a high temperature paint or enamel, such as engine block or woodstove paints rated for 1200 to 2000°F, but we more often simply use food-grade oil such as coconut or safflower oil, and cure the barrel in place like a cast-iron pan.

For decorative options, see "Barrel Decoration," in Appendix 6.

Bits and Pieces:

- Braided woodstove gasket — approx. 7 feet for each joint (14 feet to make both lid and whole barrel removable for maintenance access)
- Foil tape and/or sheet metal screws for duct joints
- Masking tape and masking materials to mark levels and protect nearby walls/floors
- Framing materials (scrap lumber, screws and driver, saw) for floor and roof work

Planning and Prep

Structure and Proportions

Critical proportions include:

- Appropriate cross-sectional flow areas to avoid constrictions
- Correct heat riser height in proportion to the firebox and flame path.
- Appropriate length for heat-exchange pipes or channels.
- Appropriate exit chimney for the building — usually a vertical chimney exiting near the roof peak

Most projects require some adjustment to fit in the actual space. Slight adjustments to the length or shape of the heat-exchange mass are often successful; changes in the combustion unit often cause performance problems. Critical proportions are described in more detail in Chapter 6.

Example heater dimensions:

(All dimensions are ID [interior dimensions] unless otherwise noted.)

- Stovepipe and channels: 8″ ID round pipe, or 7″ x 7.5″ square, = 52 square inch CSA.
- Firebox dimensions: 16″ feed tube height; 24″ burn tunnel length; 48″+ heat riser. All channels 7″ x 7.5″ in cross section.
- Firebox masonry thickness: a total *8″ minimum thickness,* includes 1.5″ to 4.5″ brick, 1″ to 2″ insulation, plus outer masonry. Total width: about 24″ OD (outside dimension) for feed-hearth area.
- Heat riser thickness: 1.25″ brick plus 1″ to 2″ insulation and insulation cage. Total width: 13″ to 15″ OD, diagonal 18″ to 20″.
- Bell/manifold area: 23″ diameter barrel, 5″ minimum masonry thickness, minimum width 33″ OD.
- Combustion area total dimensions: Total is about 3′ by 4′, plus appropriate clearances. Actual minimum footprint is 33″ wide by 45″ long. Height from footing/hearth to top of barrel is 54″ to 56.″
- Heat-exchange bench: Length 10 feet, with 22 feet of heat-exchange channel.
- Bench width/height: about 30″ wide by 18″ tall. (Two 8″ pipes, located 2″ to 4″ apart, with 5″ masonry thickness on both outer sides, gives a bench 28″ to 30″ wide and 18″ tall.)
- Exit chimney: Example has a vertical chimney about 20 feet tall, extending 3 feet above the roof, near the peak (2 feet above anything within 10 feet).

It is possible to vary some of the lengths given here. Similar heaters have worked successfully with bench lengths 8 to 15 feet long, and with 8″ ID chimneys between 8 feet and 35 feet tall. With stout foundations, you could also lay the return pipe directly above the first run, for an 18″ wide by 30″ tall heated mass (that's about the right height for a dividing wall, desk, or workbench).

Clearances, Thicknesses, Cleanouts

The figures shown here give approximate clearances and thicknesses compatible with safety, based on our experience, and our understanding of US building codes. We are reluctant to prescribe exact numbers, however, as each jurisdiction can determine its own requirements, and it's also possible to build a stove roughly to these guidelines yet produce excessive heat.

When in doubt, a few extra inches of safety clearance can make the whole project much easier and safer. See Chapter 6 for more details.

Metal Bell

The upper bell generally operates at similar temperatures to a metal box stove. I prefer to follow local guidelines for non-certified wood stove installations; generally about 18″ minimum clearance to combustibles when using a heat shield with 1″ air gap. With very excellent heat shields, minimum clearance could be 12″ from combustible walls. (See Chapter 6 for more detail on heat shields.) Adding masonry or earthen plasters to a wood-framed wall counts as a heat shield, but does not make the wall noncombustible.

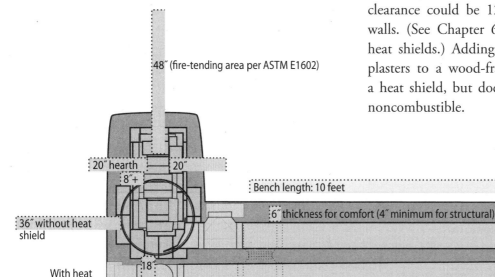

It's not that hard to get rid of a combustible wall. Either take it out (if it's not load bearing), or replace it with a masonry wall to store some extra heat from the barrel. A load-bearing 4x4 wooden post can be replaced with a metal post, or a masonry pillar or wall.

The bottom line is you need safe surface temperatures of any nearby combustibles once you're actually running the heater. That includes wood, drywall (it's held together with paper, after all), painted or varnished surfaces — just about anything that's not bare brick or stone. If any combustibles are too hot to touch (more than about 150°F/65°C), don't run the heater until you improve the heat shielding. Also, try not to trap combustibles in contact with the shielding where you can't check the temperatures. Wood or paper products that bake at low temperatures for long periods can become more flammable, releasing flammable gases and possibly igniting at temperatures as low as 200°F (95°C).

DO NOT bury the metal bell (or any part of the combustion unit) within load-bearing walls. Even if they are non-combustible, metal expands with heat and will crack surrounding masonry, and a wall is a big obstacle to inspecting or repairing the heater.

If the bench passes through a wall to heat other rooms, that wall should be non-load-bearing and noncombustible immediately around the heater. The overall design should allow excellent maintenance access to all working parts of the bench.

Firebox and Hearth

The 8″ masonry thickness around the firebox opening also acts as a hearth. Many builders include an additional hearth area of 12″ or more around the outside of this masonry, to catch any popping sparks.

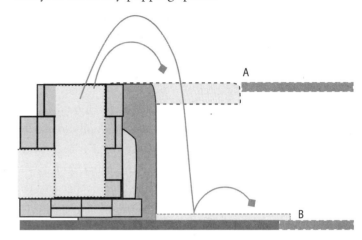

Heat-exchange Bench

If there is at least 5″ thickness of masonry around the heat-exchange channels, and 8″ around the firebox, US/Canadian code (IRC R1002) allows a minimal clearance from masonry surface to combustible walls of a 4″ air gap. Otherwise, it would require 36″ clearance. Air must be able to move and rise freely in this gap, carrying heat away from combustible structures.

Since the bench temperatures typically remain low enough for skin contact or cushions (85° to 110°F), some builders just build directly up against a wall. The 4″ is a good practical minimum however: in addition to meeting code, it allows you to reach inside to finish, maintain, check temperatures, and clean any fallen debris or intruding vermin; it reduces heat transfer to walls — especially exterior ones; and is a good width to create a steady flow of warm air for circulation to other rooms.

If you don't have combustibles nearby to worry about, the practical structural minimum for the masonry around the ductwork is about 4" (one width of brick, or a handspan of cob). This prevents lateral (outward) forces from cracking the masonry around the pipes, and produces comfortable surface temperatures in most climates after 2 to 6 hours of firing.

We prefer to build our benches thicker around the first (hottest) section — about 5" to 6" thick — then gradually slope the duct toward the surface to give a 4" or 5" masonry thickness at the other end of the bench. Burying the hot pipe deeper and the cooler pipe less deep gives more even surface temperatures for comfort, and also helps with gas flow.

Exit Chimney

High temperature (HT) chimney installations are generally at least 9" from combustible walls (with heat shielding or double-walled chimney in that area); follow manufacturer's instructions. Class A insulated chimney sections are used for tighter clearances such as the ceiling and roof penetrations. Be sure any insulated or double-walled chimney sections have *internal* dimensions to match your system size: 8" ID (inside dimension), not OD (outside dimension).

Even though rocket mass heaters often have a low-temperature (LT) exhaust, we prefer to use high-temperature (HT) chimneys as for other masonry heaters and woodstoves. We have seen many owners repurpose old chimneys for new woodstoves, which can produce unexpected creosote from burning sponge-wet wood, or the need to get terrifyingly creative with chimney-priming methods. A high-temperature (HT) exhaust chimney offers peace of mind to both occupants and building officials.

There are huge safety and performance differences between a proper, best-practice, to-code chimney, and a chimney improvised or "improved" by an inexperienced installer. Essential chimney functions include vertical rise, warmth, heat tolerance, tolerance for corrosive and damp exhaust gases, airtight seals, screening to stop sparks and vermin, and weather protection. See Chapter 6 for more details.

Cleanout Access

We put cleanout access at each 180-degree turn, and at the bottom of each vertical drop, including the manifold. The goal is to give access to every part of the system with available tools.

Allow at least 1.5" open around each cleanout cap for ease of removal, or add a handle. Sometimes metal cleanout covers can get too hot to touch; give them extra safety clearance where possible (it also makes it easier to get into them when cleaning). You can cover the cleanouts with insulation-backed tile, or with 5" of removable masonry (such as a fitted tile cover with loose brick behind it) in order to meet masonry heater codes for reduced clearances.

Location

Rocket mass heaters are a radiant heat source, and work best by direct contact or line of sight. Therefore, the optimal location is in the occupied core of the house.

In real-life versions of our example project, the owners defined this core location in different ways. In one case, it was in the living room of a log cabin. Another demonstration project serves as a warm divider in a multi-use barn. Several similar projects collect heat from the south-facing windows of sun rooms which double as winter hangout space.

Foundations play a major part in deciding where to locate the heater. In the case of one sunroom, the existing patio slab offered a better foundation than the manufactured home, simplifying the project. In another sunroom with suspended floor, the owners cut a new opening, reframed the floor, and built up a noncombustible heater foundation using slab and cement block. The cabin and barn sites had compacted gravel or dry, undisturbed grade throughout, so the location was based on optimal heating of the space.

The flat bench in this example presented in this chapter can serve as a bed or full-body heating pad. With the heater across from a sunny window, it stores solar gain and saves fuel.

The bell is near the center of the space, in line of sight of several room doors. Any room from which you can see the bell is a room which it can heat.

See Chapters 2 and 3 for more details on design, Appendix 3 for special situations, and Appendix 4 for general home design.

Site Prep and Cleanup

Construction projects are messy. With masonry, you will have mud, and possibly dust. You can avoid having to breathe masonry dust by keeping the mud damp throughout the building process, and/or wearing good-quality respiratory protection and ventilating the space.

Remove all valuable or delicate items, and use plastic or fabric sheeting to cover

> The most popular place to install a rocket mass heater is along one wall of a living room or great room (usually replacing a sofa).
>
> The second most popular place is wherever an existing chimney, concrete slab, or new addition makes installation cheaper or easier.
>
> Other popular options include:
>
> - a heated guest bedroom or reading nook
> - a heated divider between living room and kitchen (or office)
> - a heated floor or sub-floor for flexible space
> - a new bay window or nook addition off the main room
> - as thermal mass in a solarium or passive solar design

anything that can't be moved. Sweep floors well before covering with clean tarps or painter's dropcloths, and tape the edges down. (Grit under a tarp can be worse than no tarp.) A cardboard or plywood path in high-traffic areas protects floors from gritty wheelbarrows and muddy feet.

Just outside the project space, designate staging areas for each type of material, and a wash station for cleaning tools and hands. Never rinse mortar, clay, or any masonry materials down the sink: they block up the plumbing. A rinse bucket for pre-washing saves a lot of water and reduces puddles from hosing off.

Practice Before Starting

Dry-stack the firebox (dry = without mortar), either in place or outdoors. Draw up a full-size template showing finished dimensions, and check the clearances in the actual space. Lay out the ducting and check the chimney connection. Practice with any unfamiliar tools and materials, including site-sourced materials. If using earthen masonry, make test bricks as described in Appendix 1 and allow them to dry. Mock up the bench in the space with planks or buckets to try out the intended seat height.

Laying Out the System

We often lay out the system, or mock it up in the space, before beginning any permanent alterations.

The first thing to do is to check rafter and joist spacing in the attic to ensure your plans are feasible. For a plumb exit chimney and minimal cutting of pipe, you will install the through-roof chimney first, then connect the rest of the system ducting working backwards from this point.

Once you are sure you can install the chimney where you want it, do so. If the chimney location changes, recheck clearances and other project dimensions to make sure everything will still work. Lay out all pipes, connected, paying attention to the leveling. Use noncombustible rubble such as pieces of brick or rock to prop the horizontal pipes in their proper places.

Aim for a slight, continuous upward rise from heat source to chimney (see diagram, "Thickness variation"). The rise helps exhaust move smoothly along the pipe, and prevents hot spots and condensation pooling.

A slope of about ¼" over ten feet is sufficient — just don't let it dip back down anywhere. Pay special attention to elbows — if they are not exactly 90 degrees you may need a steeper slope to keep them from making little lumps where hot exhaust can pool.

Always keep the intended finish height in mind. You don't want the pipes to end up too high. It's always possible to add more tile or cushions, but impractical and possibly dangerous to make the bench thinner than 4" over the highest pipe.

Mark the wall at sub-finish level with a chalk line and/or masking tape. A line level for a chalk line, or a cabinet installer's laser level is lovely here. If you don't have one, a long bubble level, or a long board with any bubble level, will get it done.

In our role as trainers for DIY and novice builders, we don't draw the "finish" line at first. This tends to result in ducts, rocks, or other lumpy materials being laid up too high. Instead, we clearly mark the height of our current task only. First we mark the ducting height, and lay and build up to that level. Then we mark the infill or sub-finish height, and build to that level. We wait to mark the final finished height on the wall until we are ready to work with finish materials such as tile or plasters.

Mock up the brick firebox. Note the actual achievable dimensions in case your brick is a different size than shown. Make a cardboard template if you like, showing actual dimensions using your bricks and intended mortar thickness. Balance the barrel in its intended location. Check clearances to combustibles (allowing for 8" of masonry around the firebox, 5" around the barrel and pipes).

Once you are comfortable with the spacing and layout of the heater, mark intended dimensions on the floor and on a removable template or plan. You can move the parts out of your way (leaving pipes connected in long sections if you like), and begin the permanent installation.

Footings/Foundations

Foundation requirements vary from place to place. Earthquakes, frost heave, expansive soils, local laws, existing structures, etc., may dictate your foundation type. See Chapter 6 for more detail. A typical footing might be a 4″ or 6″ slab on grade.

With cob projects (monolithic earthen masonry), we almost always include a dry stone footing right under the cob, even on projects with a concrete slab foundation. It's how cob was built traditionally, with emphasis on good drying conditions rather than moisture barriers (which can become condensation traps).

Preparing Foundations Over an Existing Basement or Crawl Space

If you are building your heater over an empty space like a basement or a crawl space, you will have to take some extra steps.

1) Determine heater placement and size. Inspect for hidden floor beams, rafters and joists in the through-roof area, and other factors that may affect heater placement. Calculate the total weight of slabs, concrete block, brick, or other foundation materials as well as the heater, and consult code or engineering tables for appropriate foundation thickness and reinforcement.
2) For foundation size, add at least 6″ all sides, and 20″ hearth.
3) Brace up the floor from below, and block the joists in place.
4) Cut a hole through the floor and joists. The new framing should allow a 2″ gap between framing and the sides of the heater.

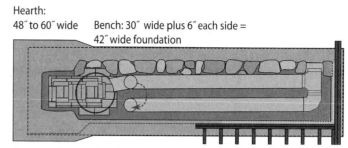

Hearth: 48″ to 60″ wide   Bench: 30″ wide plus 6″ each side = 42″ wide foundation

Framing: 2″ gap each side = 34″ opening around 30″ bench

Example Loads and Footings

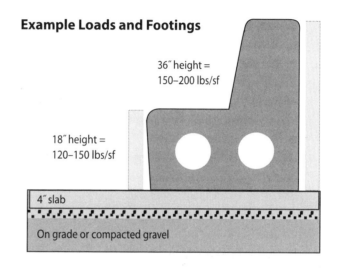

36″ height = 150–200 lbs/sf

18″ height = 120–150 lbs/sf

4″ slab

On grade or compacted gravel

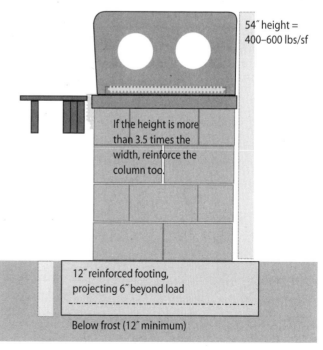

54″ height = 400–600 lbs/sf

If the height is more than 3.5 times the width, reinforce the column too.

12″ reinforced footing, projecting 6″ beyond load

Below frost (12″ minimum)

5) Reframe the floor around the heater opening, with double or triple headers as required by floor loads, and 2" gaps between framing and planned masonry outline.

6) Build formwork, and pour the necessary foundations.

7) Build up to the desired height using non-combustible supports (concrete/block/brick).

8) Consider including insulation in or below the pour (e.g. as perlite aggregate in concrete above grade, or insulation fill in cinderblocks), and beside exposed walls (e.g. the footing walls within the crawl space).

9) After building the heater, use noncombustible insulation and trim (e.g. tiles as for hearth) to bridge the gap between footings and suspended wood floors.

Exit Chimney

Our sample heater has a vertical through-roof exhaust with 20 feet of vertical chimney. We used a manufactured Class A chimney, as for any woodstove or masonry heater.

The chimney must fit the building as well as the heater. It should be the tallest element of the house, 2 or 3 feet above anything within 10 feet (such as the roof ridge). Similar heaters have been used successfully with chimneys from 6′ tall to about 30′ tall.

An existing chimney can be used, as long as it is in good condition and the flue interior is the right size for the heater (8" ID round, or about 50 to 55 square inches if rectangular).

Chimney Methods: Up or Sideways?

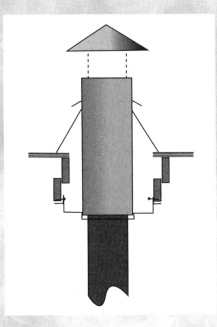

Most combustion devices rely on the exit chimney for their primary draft. A warm, tall exit chimney keeps the fire flowing properly. Backdrafts can happen when an exposed outdoor chimney is cold, or when any combination of winds and house drafts result in higher pressure outside the chimney than in the room.

Experienced installers locate chimneys near the peak of the roof, with most of their height inside the house. (For some excellent information about chimney design, see www.woodheat.org.) These sheltered chimneys stay warm, share more warmth with the house, are more efficient and easier to operate. Tall, sheltered chimneys escape most ill effects of wind gusts and ordinary household pressures. And they usually cost less than an exterior chimney, once both are insulated properly.

The building code (IRC R1003.9) describes chimney termination height as "at least 2 feet (610 mm) higher than any portion of a building within 10 feet (3048 mm), but shall not be less than

3 feet (914 mm) above the highest point where the chimney passes through the roof."

Ianto Evans's original rocket mass heaters often used a side exit like a dryer vent. Unlike box-style woodstoves, rocket mass heaters with no adverse pressure effects may draw even without an exit chimney, thanks to the internal heat riser. But Ianto's own building designs are pretty unusual. As described in *The Hand-Sculpted House*, his tiny Cob Cottage houses are compact, one-story, earthen masonry cottages with membrane-lined living roofs. These buildings have noncombustible walls, no roof vents, and most are in sheltered woodland settings without extreme winter temperatures. The house acts more like an oven or hat, less like a chimney.

In taller homes with roof vents or upstairs windows, the house itself acts like a chimney. Warm air rises and pushes out the top; cool replacement air flows in at the bottom. Wind gusts can cause back-pressure on any side of the house, and especially around corners and eaves. In these ordinary conditions, the chimney opening must be higher than any other holes in the house — and above any corners such as eaves and ridge — to out-compete house draft pressures.

For conventional homes, wall exits may seem cheaper but usually aren't. They require more insulation ($$$), and may need special priming (preheating), supports, and water protection (from water running along pipe into wall, and from ice dams at eaves which can rip a chimney away from the house).

If a wall exit seems necessary, put it as high as possible on the gable (end) wall. The chimney will need clever bracing and supports to reach above the roof peak, as well as protection from pressure, wind gusts, vermin, and weather. (More details in Appendix 3.)

We still experiment with horizontal exhausts occasionally, when the owners insist, because an ultra-low

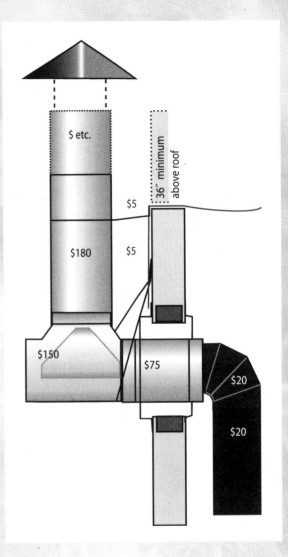

temperature exhaust is a seductive dream (low exhaust temperatures imply less heat is being wasted). We have learned to include a capped T inside the building so the owner can install a vertical exhaust later. In 90% of modern buildings, a conventional chimney is the simplest *reliable* solution.

Notes before doing your own chimney installation:

Take to the Heights: Installing higher up in the roof provides better draft, is easier to support using readily-available ceiling brackets and through-roof kits, usually costs less because it takes fewer insulated sections, and involves less overall risk of leaks.

If your house has several roof heights, the chimney should be at least as tall as the highest roof. If forced to install near eaves, you still need to raise the chimney to the same final height above the roof with insulated sections, and you will need to provide more structural support to secure the longer, exposed chimney sections against wind and weather. Don't forget that there is effectively a lot more weather to deal with near the eaves, since the roof is shedding water and ice in that direction. In snowy climates, chimneys down near the eaves will catch big loads of snow and ice, which can rip the chimney away if they are not effectively diverted by a protective structure such as a blind dormer or "ice knife."

This End Up: Chimney sections are installed male-ends-down, female-ends-up, to catch any dribbling rainwater or creosote-tinted condensation. (Gas and smoke don't care about seam direction. Focus on keeping any unsightly liquids inside the pipes.)

The ceiling collar, chimney cap, and other parts are all designed for this orientation, and may be difficult to fit together if the sections are installed upside down.

INSTALLING A THROUGH-ROOF KIT (EXAMPLE)

Each chimney kit comes with its own instructions. Follow the manufacturer's instructions. Here's a sample installation, using a common type of roof kit for asphalt or shingle roofing.

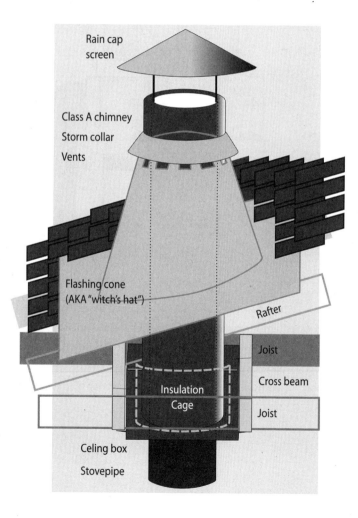

1. Measure your roof slope and the distances from your project to the ceiling, through the attic, and above the roof. Purchase a through-roof kit suitable for your type of roof and ceiling.

    Slopes are measured as rise-over-run, such as 8:12 or 10:12. A roof angle such as 30 degrees would be a 1:2 or 6:12 slope and 45 degrees would be a 1:1 or 12:12 slope. Get matching parts, including any necessary sections of insulated chimney, a chimney cap, and anything else they didn't include in the kit.

If shopkeepers ask for a model number, you can tell them you are installing a site-built masonry heater requiring 8″ ID stovepipe and Class A HT (high-temp) chimney.

2. Mark the desired through-roof location with a probe through the ceiling, 9″ from the wall. (Poke through with an icepick or long screwdriver with wood or rubber handle). Don't stab any electric wires!

From the attic, compare the probe location with nearby utilities, joists, and rafters. Center the hole a safe distance between obstacles. Avoid any location that could trap water in roof valleys, or cause undue difficulties with through-roof installation, wiring, or other utilities.

3. Drop a new probe from the attic to mark the intended chimney location. With plumb line or level, compare this location to the heater plan. Check location, thicknesses, and clearances for chimney and heater.

4. If there is no clear path between framing, you can cut and reframe a rafter or joist. (See lower diagram in sidebar, "Roof Framing"). Do not cut manufactured trusses or ridge beams. Locate the chimney to one side instead.

5. After setting the final placement, follow the directions on the roof kit for installation. In the top example of "Roof Framing," we show how we added cross-bracing (pale lumber), cut the ceiling between joists to fit the ceiling box, and then attached the ceiling box to the joists and bracing.

6. Use a level or plumb line to mark and cut a rough hole in the roof. Cut the hole at least 2″ bigger than the insulated

Roof Framing

Center stovepipe between existing rafters and joists. If necessary for adequate clearances, cut and reframe one rafter (shown) or one joist. Avoid cutting ridgebeam or trusses.

Roof framing:

Center pipe between existing rafters and joists.

If necessary, for adequate clearance, cut and reframe one rafter (shown), or one joist. Avoid cutting ridgebeam or trusses.

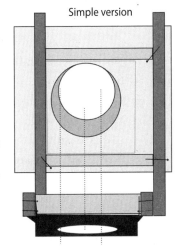

Simple version

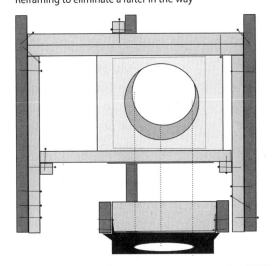

Reframing to eliminate a rafter in the way

pipe, yet smaller than the base of the cone-shaped flashing.

7. Connect appropriate chimney sections through the ceiling box, male-end-down (keeps liquid condensation inside the pipe, not drooling down the outside). In most cases, you need insulated Class A chimney sections for the ceiling and roof penetrations, and single- or doublewall stovepipe in heated rooms. Keep attic insulation away from the pipe. Kits may include an insulation shield for this purpose.

8. Line up the chimney section(s) vertically through attic and roof, then fit the flashing/boot around the protruding chimney section.

9. Fit the flashing over the shingles below, and under the shingles above. For best results with some roofs, especially if the shingles are easily damaged or hard to make leak-proof after disturbing them, you may need to re-shingle from the chimney up to the ridge.

   Note: This is another reason to install the chimney near the ridge line, if possible: it makes leak-free installation much easier. It also tends to reduce future problems with wind gusts, ice and snow loads, and maintenance.

10. Use goop from the kit as directed to seal flashing and roof. (Try not to create dams that will cause water ponding; smooth any exposed caulk to let water run off easily.) Seal the storm collar around the chimney.

11. Add insulated sections of the same type to extend the chimney to the proper height above the roof (3 feet above the roof, and 2 feet above anything within 10 feet — including snow loads). Use bracing or guy wires to support the chimney if it extends upward more than about 5 feet above the roof.

12. Install spark arrestor/screen, and chimney cap, protecting the chimney opening from gusts blasting directly down the chimney, sparks, and vermin. Allow cross-flow from any direction for best draft, and make sure the opening and screen size are big enough for 100% of the system cross-sectional area (CSA).

For chimneys near other tall buildings or in gusty climates, consider wind blockers, pivoting "weathervane" or "turbine" chimney caps, or H-shaped pipes made from 3 T's (as seen on boat stoves).

Through-wall chimneys are not ideal. (See sidebar, "Chimney Methods: Up or Sideways?" and Appendices 3 and 4.) However, if a through-wall chimney seems like the best option, put the exit as high as possible on a gable wall or near the roof peak. You may need two or more storm collars to stop rain running down the chimney into the wall. As with all manufactured chimneys, the through-wall should be within 30 degrees of vertical if possible, which means the hole will be a relatively long ellipse. You will need at least one insulated adjustable elbow or 30-degree elbow; the other elbow may be single-wall if it's at least 9" from combustible walls and ceilings. Extend insulated chimney to the proper height above the roof (Step 11 above), and use a suitable chimney cap (Step 12).

Note that larger number of insulated sections generally makes through-wall chimneys more expensive than the standard through-roof — and they perform worse.

Combustion Unit

The combustion unit is the hottest area of the heater. Correct proportions, insulation and expansion jointing, and double-sealing to prevent air leaks are critical for good performance.

The dimensional goals are:

- a 50-square-inch flow area (7" x 7.5" ID channels)
- a 16"-tall vertical wood feed (no taller)
- a burn tunnel about 24" long (give or take mortar thickness)
- a heat riser at least 48" tall

Note that the burn tunnel length may need to be adjusted for any mortar used, or if using a different dimension of bricks. (For example, 4 bricks plus 3 mortar joins + 14" (two 7" openings) — with ⅛" mortar joins and 2.5" firebrick on edge, the new length may be 24⅜" instead of 24.") Alterations less than ½" do not usually affect performance, as long as walls remain flush and plumb.

Brick Masonry
Brick Base Pad

- Place 1" to 2" of sturdy, noncombustible insulation below the brick floor pad (we use clay-stabilized perlite, but acceptable alternatives include refractory insulation board, or a layer of insulating brick).
- When leveling is needed, use the clay-stabilized perlite insulation, or a layer of sand or mortar, to level the brick pad. Use screed boards (thin dimensional lumber) to quickly create a rough level. Do not compact the material yet.
- Lay the bricks to make the floor pad, but do not press down. Once all bricks are in

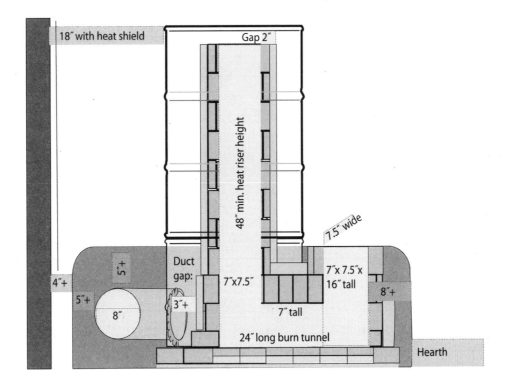

place, use a level to find the proud (too-high) bricks. Tap the highest bricks down to level, using a wood or rubber mallet, or the wooden handle of any hand-tool. Also tap bricks inward horizontally, so they sit firmly against each other.
- Once the whole pad is level, brush sand between the bricks to stabilize them. Use mortar or cob to support bricks from the side.
- *Do not allow anyone step on your level pad,* as it will undo all your work.

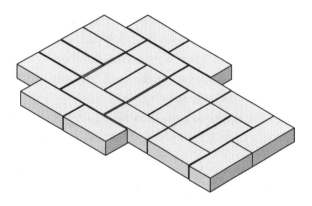

Firebox

The lower part of the firebox is all one big rectangle, the bottom of the burn tunnel. Above this space will be the future openings for feed tube and heat riser, and the brick "bridge" that separates them. Standard interior dimensions for this box are 7.5" wide by 24" long.

Mark the inside shape of this box on the level pad, using your template again to check placement.

Do You Need Mortar?

Traditional mortar is not brick glue; it is "liquid shims" for raising each course of irregular brick to a true level. Good masonry work supports itself under gravity. Mortar just keeps uneven courses from wobbling.

New firebrick is supposed to be perfectly dimensional and square, for stacking without mortar. If there are no irregularities in the bricks or pad, you can use a thin-set (we use clay slip, a slurry of fireclay and water) instead of a gap-filling mortar.

If you (or local authorities) prefer a hard-bonding refractory, choose a thin-set refractory product rated for 2400°F or higher. We prefer clay slip for thermal performance, price, ease of work, and ease of repairs.

These directions will continue as for clay slip. If using a different product, follow the product's instructions for mixing, set time, curing time, etc.

If either bricks or pad are irregular, mortar may be needed to fill any gaps and keep your courses level. Your mortar join should be as thin as practical; screen the ingredients before mixing to eliminate over-sized grit. Choose a thickness, such as ⅛", and practice to achieve it consistently.

See Appendix 1 for more detail on traditional clay-based construction and modern refractory options.

Preparing Mortars and/or Clay Slip

Clay Materials

Clay slip is a slurry of water and clay, just thick enough to dimple when splashed — like a thick paint or runny pancake batter. Good slip should coat fingers thick enough that you can't see fingerprints. Ten to 25 lbs of dry clay, or ⅓ to ½ a bucket of damp clay, will make about a 5-gallon bucket of clay slip. Grate any lumpy clay through a wire mesh screen (½″ or ¼″ hardware cloth works well) to screen out rocks and speed the dissolving process. Mix with hands, a large whisk, or a paint-mixer or stucco-paddle on an electric drill.

Clay-sand mortar may be 3 to 5 parts sand to 1 part clay slip by volume. Mix thoroughly, and add water if needed until a ¼″ layer will just support a brick's weight on its own, but goosh out if you push or tap on the brick.

Pure clay products have no "curing time"; they harden as they dry, and soften again if re-hydrated. It's the water and clay together that make the bond. Pre-soaking the clay to make sure it's fully hydrated often results in easier mixing and a better-quality material.

Refractory Cements and Mortars

Follow the product instructions, including how to thin or prepare the mix, and pay careful attention to working times (not the same as curing time). Prepare modest amounts — no more than you can use immediately, and discard any material that begins to set before use. (Some products are sensitive enough that you can't mix new material in cups that have older, partially-set material. Have a good selection of disposable measuring cups and mixing boards available: you won't have time to clean them for re-use while trying to lay level courses.)

In the instructions below, clay slip and "thin-set" refractory cement products are used the same way, as a runny coating. "Mortar" can refer to either clay-sand or refractory mortars, thicker materials that can fill uneven gaps. If using refractory mortars, pay special attention to curing times, and you may need to add expansion joints

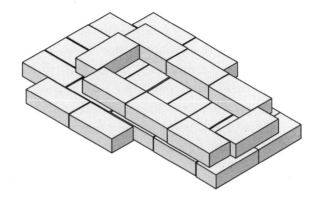

which would not be needed with earthen materials.

Laying the brick: Before applying clay slip or mortar, wet each brick in a bucket of water. Dry bricks won't bond well with clay, and can suck water out of refractory mortars and prevent them from curing properly.

Clay slip or thin-set method: Dip each brick in the thin-set or slip, or paint the edges of the brick, just before you set each brick.

As you complete each course, tap any high bricks slightly to bring them level. (You will not be able to correct big problems like you could with mortar, so keep things tidy as you work.)

Mortar method: If your bricks are uneven, you may need a thicker mortar to shim them into place.

- Apply the mortar in an even layer or ridge under the first course of bricks. Make the ridge slightly taller than your final join thickness, so you can tap down the bricks.
- When you set the first brick, align it slightly inside the inside corner, with room for mortar.
- "Butter" the end of the next brick with a ridge of mortar, and butt it up against the

first brick. Orient each ridge to provide an airtight seal — parallel with the joints that will be visible from outside the wall/box.

- Hold the first brick steady, and tap the second brick into place with a wooden block or rubber mallet. The ridges of mortar should goosh as you place the bricks, establishing a good seal, and allowing the brick to shift into place as you tap.
- Butter each brick just before you lay it; this keeps the mortar fresh and workable, and makes it easier to butter the correct side of the brick.
- Level each course while the mortar is still fresh, before beginning the next course. Use a level to check before tapping down. Sometimes a brick tilts one way yet the course does the opposite. Tap high bricks into place. Avoid pushing, wiggling, or wobbling the bricks — it breaks up the mortar seal. If the brick doesn't tap down far enough, you may need to clean it and reset it with fresh mortar.
- When each course is done, smooth any scant or bulging mortar joints with a trowel or wet finger.

All methods:

- Start at one inside corner. Ignore the outside corners; they won't match up. Work each course in a spiral, butting the inside corners against the established bricks. (The spiral layout lets each side have a free edge. Using this method, we can make precise interior dimensions without cutting bricks. The outside edges won't usually line up. We generally ignore them, unless clearances are tight enough to warrant the extra work of trimming bricks.)
- Each course must be offset to avoid running gaps. Isolated, vertical stacks of material tend to tip and fall (like Legos). In all but two places in this layout, a solid brick bridges over each joint in the previous course. The layouts show how to achieve this using a spiral pattern. One course is laid clockwise, and the next counter-clockwise.
- The bridge requires two bricks cut to 7″ length. (The edges of these bricks will line up with a gap in the course above or below, making a slight running join.)
- The cleanest way to cut brick is with a tile-saw or diamond-blade masonry saw. Mark the cut, then secure the brick on a cutting surface. Scrap bricks can be used as blocks if needed. With water running on the masonry saw blade, make the cut patiently and smoothly.

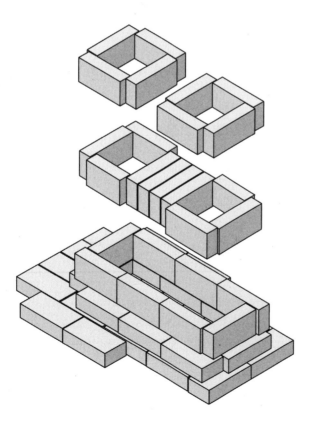

- If you do not have a masonry wet-saw on site, you can also use a grinder or cold-chisel, and sand down any rough spots on a convenient patch of sidewalk. Mortar can fill any little gaps, but closer is better.
- Sometimes the bridge bricks do not precisely fit the gap between heat riser and feed openings. Do not make the heat riser smaller, or the feed opening larger. The feed must be smooth and flush to prevent wood hanging up on protruding ledges. If adjustment is necessary, either cut a thinner brick for the back of the bridge course, or you can tolerate a small discrepancy at the back (heat riser) end in this course only.

Heat Riser:

- Lay the heat riser much the same as the first two courses of firebrick, making spirals of brick and paying special attention to internal dimensions, level, and plumb.

Use just enough mortar to level each course. If the mortar is too thick or wet, you may find that lower courses goosh out when you stack the next course on top.

Working the Sunwheel (general method for working to inside dimensions):

Start by guessing one inside corner. The next brick butts up against the previous brick (don't force yourself to guess any more corners). As you work, use your fingertips to feel inside for alignment with the course below. Continue working the same direction until all four sides are finished. Check for level. Tap down high bricks or build up low areas with a small amount of mortar, if needed. Then begin the next course, alternating clockwise and counter-clockwise to stabilize the corners.

Level the whole stack, then take a break until the mortar sets.
- Our sample heat riser is half-firebrick (1.25" × 2.5" × 4.5"). Alternate courses clockwise and counterclockwise to offset the joints. Check each course while the mortar is still fresh. If a course will not level with thin-set, add fine sand and make a just-thick-enough mortar layer to make up the level.
- Alternative: A pre-cast heat riser can be substituted for the bricks above the bridge course. Set on a ½" ridge of fresh mortar. Level and plumb, tapping the high side down to correct any problems. Smooth the mortar around the join for a good seal, then allow to set.

Insulation

We insulate around the heat riser and firebox with refractory insulation. For this project, we use ceramic-fiber refractory blanket, available at fireplace suppliers, but many builders have also used a metal cage filled with perlite or vermiculite. Light, crushable insulation materials also serve as an expansion joint around the firebox to protect outer masonry.

We carefully wrap the insulation blanket around the brick heat riser. Good air seals are as critical as the insulation in this area, so we try not to bump the bricks too hard

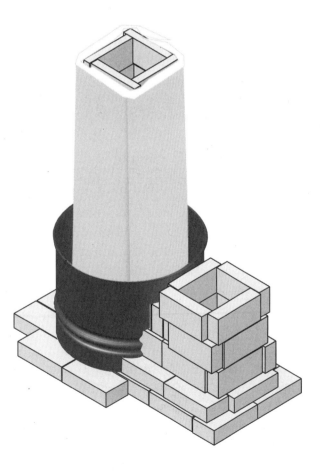

while we work. We butt or lap the edges of the insulation, and wrap a wire around it to hold it in place for the next step.

Around the insulation, we wrap and wire together a cage of ¼″ mesh (to protect the insulation during cleaning). The lower section of mesh (and blanket) must be cut to fit over the bricks of the burn tunnel and bridge. We use the scrap pieces to insulate the top of the burn tunnel.

As an alternative, clay-stabilized perlite is easy to use under and around the burn tunnel. To hold it in place around the heat riser, we sometimes make a cage of sheet metal, or fine wire screen, and pour in the same perlite mixture as we used below the brick pad. (Galvanized material survives just fine in this area outside the insulation, up to about 4″ below the heat riser's top.)

Clay-stabilized perlite is made by coating perlite with clay slip. Mix thoroughly to create a material that will clump together, but is still very lightweight. This material can be packed in self-supporting shapes up to about 12″ tall, or supported with metal or brick courses.

Suitable alternatives: The following insulations may be substituted for refractory blanket:

- Insulating brick (kiln brick or refractory light brick), 2″ to 3″. If this brick is used alone, without an interior lining of dense brick, it should be rated for 2500°F or better. (Due in part to its excellent insulating qualities, the insulating brick may get hotter than 2300°F on its inner surface. Dense brick conducts some heat away from the surface, and may stay at slightly lower temperatures overall — we've seen ordinary clay brick, the red building kind that should not be rated much above 2000°F, handle this job time after time.)
- Ceramic-fiber blanket or board, 1″ refractory insulation (usually rated for 2100°F or better)
- 2″ rock-wool or high-temperature board or batt insulation (various products are rated for 1400 to 2150°F; do not use fiberglass, which completely melts around 1300°F)
- 2″ of loose or clay-stabilized perlite (perlite melts around 2300°F, the clay does not need to be rated), in wire cage or sheet-metal container
- 4″ of loose vermiculite (no clay) in metal container (handles over 2000°F)

Not suitable for insulating heat riser:

- Pumice or porous lava rock — you would need about 9″ thickness, which does not fit in available space. Expanded-clay pellets and loose dirt are also too dense, so not suitable.
- Straw-cob is too heavy, and light straw clay can burn out and lose structural integrity at the operating temperatures around the heat riser. Some people have reported partial success making ceramic insulation using sawdust and clay, or coal-dust and clay. (Search for "vernacular insulative ceramic" online.)
- Fiberglass often melts at typical operating temperatures. (Combustible insulation such as wool or cellulose is also unsuitable, obviously.)

Note About Building Sequence

The thermal mass and manifold jobs can be done simultaneously if there are enough people on the team. If only one or two of us are doing the job, we typically use the following sequence:

- Lay one or two courses of masonry below the pipes, avoiding the manifold area.
- Shim up the pipes to their proper placement. We prefer to use noncombustible masonry shims (wedge-shaped rocks, spare bricks, or stiff mortar) so that there are no combustible supports that will need to be removed later.
- Fabricate and seal in the manifold.
- Go back and build the rest of the thermal mass around the pipes and manifold.

Metal Radiant Bell and Manifold

The manifold in our sample design is a cut-off barrel, forming an open cylinder that supports the upper barrel. The height of this partial barrel is adjusted to actual measurements before cutting and fitting.

If you do not want to use a second barrel, it is also possible to build the manifold from masonry.

The upper barrel needs a sealed top and open bottom. We generally use the removable top lid as the barrel top, and cut out the bottom, for ease of cleaning. Other types of barrels (or a damaged surface) may dictate which end is cut away and which end becomes the top.

All metal parts must be clean of rust and debris. For bell and metal manifold parts reclaimed from scrap, any ordinary paint or varnish must be removed (by sand-blasting or burning in a hot outdoor fire). The "pocket rocket" described in Evans/Jackson (and reproduced here in the sidebar) is an effective way to burn off the paint with minimal toxic smoke.

After cleaning, the barrel may be coated with a high-heat cooking oil, or painted with a high-heat enamel rated for 1200°F or hotter, to discourage rust.

Once you have your barrel:

- Using plan drawings as a guide, verify actual measurements. The simplest and

A metal manifold may be cut to fit around the firebox (top) or a masonry manifold can be built (brick and mortar example, below).

Pocket Rocket

The "pocket rocket" can help get all the paint off a reclaimed metal barrel. Not only does it clean a barrel of paint, but it is a proof of concept for a downward-fed fire.

To minimize toxic smoke, we first wrap the barrel in damp insulation (clay-dipped paper works especially well) and then prop it up on bricks in a wide, fire-safe area. Most of the insulation will char and fall off while burning, so 5 or 6 feet of clear space on all sides of the barrel is ideal.

From an extra lid or piece of sheet metal, make a cover with a stovepipe fuel feed going down into the barrel and a chimney about the same size coming up and out. The fuel feed should come within about 3 or 4 inches of the bottom; this forces the heat down, makes the fire blast harder, and reduces smoke output.

Light a fire in the bottom of the feed tube, and if needed, blow down the tube to start the draft going the proper direction (down the feed and up the chimney). A few sticks of wood should get the barrel hot enough to burn away the paint; you'll know it's done when the insulation starts to burn through and fall away, revealing red-hot metal beneath. Once it cools, we use a scrub brush or handfuls of wet sand to remove the remaining clay and any powdery paint residue.

The term "pocket rocket" and this general method were first documented in the Evans/Jackson book *Rocket Mass Heaters*.

most accurate method, if space allows, is to make a 2" spacer from wood or brick, and carefully suspend the upper barrel on the finished heat riser rim. Measure the remaining height to the brick pad, and cut the lower barrel to make up this height. You can also calculate the height needed using a tape measure, as long as you are careful to measure from the inside of the barrel lid (not the rim).

We prefer to use the original bottom of the manifold barrel as its new top surface, for two reasons: first, the rims fit better this way. The bottom rims of most barrels are slightly smaller than the top rims, allowing two rims to fit into the same band-clamp that was used to hold the lid on one barrel. Second, we can use extra material from the bottom of the barrel to gain a more secure fit. When you cut out the bottom, make the cut about 2″ from the inside of the rim, and hammer the remaining material up to create an inside lip to support and guide the top barrel.

(As an alternative to using a metal manifold liner, you can form an airtight, hollow base of the appropriate height using brick, adobe, or cob.)

- Using actual firebox exterior dimensions, cut the manifold to fit around the terraced bricks. Allow up to ¼″ extra clearance rather than disturb the mortared bricks. Test-fit the manifold gently; re-cut where needed for an easy fit. Insulation blanket, woodstove-door gasket, and/or mortar can be used to fill any too-large gaps.
- Mark and cut the ducting hole, and a cleanout port if desired. Tab and fit 8″ ducting to these holes, minimizing any protruding edges. Using an 8″ T as the first piece provides cleanout access with minimal cutting work. (If building a masonry manifold, this T can be embedded in mortar between courses.)
- When the manifold barrel is ready to place, test-fit with upper barrel again to confirm the proper 2″ gap above finished heat riser. If lower barrel is too short (e.g. actual heat riser is taller than planned due to mortar), use masonry shims to raise up the manifold and barrel to the proper height. If lower barrel is too tall, re-cut as needed, or build up heat riser rim slightly using mortar or cut brick pieces. This is a good time to make a sloping lip on the heat riser, to shed fly ash.
- When fit is correct, mark the placement on the brick pad. Paint the pad and metal surfaces with clay slip where they will meet. Lay about 2″ of mortar bedding on the brick pad to receive the manifold piece. Without pushing down yet, slide the cut barrel in place, fitting all sides carefully.
- Level and plumb, tapping the high side down to correct any problems. If shims must be adjusted, make sure to reseal the mortar (reset with fresh mortar if necessary).
- Smooth the mortar up against the sides like caulking. Add up to 5″ of mortar or a course of masonry around the manifold base at this point if available.
- For the upper barrel with removable lid, install gasket in lid (between lid and rim), and clamp lid on with band-clamp. All materials in this area should be heat tolerant to above 800°F.
- To make a removable upper barrel, join the two barrels with the band-clamp from the second lid, with a woven woodstove gasket tucked inside. Make sure barrels are seated in perfect alignment before clamping. Foil tape may be used at this lower join (working temperatures 250–300°F) to temporarily seal the barrels before clamping.

For permanent barrel placement, seal barrels by welding, bolting, or with clamp and gasket, as above.

(Alternative: If using the masonry method, set the barrel to its proper height with

mortar and/or shims, then bed in earthen masonry 2 to 4″ above the barrel's lower rim.)
- While earthen masonry is still pliable, test-fire the system to make sure that the proportions are working as expected, and to create the thermal expansion tolerance around the embedded metal rim.

If the system will not be test-fired until later due to refractory curing or other reasons, you may create expansion joints now by wrapping the entire firebox and the metal manifold in scraps of refractory insulation, woven fiberglass gasket or ribbon, or a similar flexible, heat-resistant material at least 1/8″ thick. Expansion cracking can also be patched later, by filling the cracks (at their widest) while hot, but this patching process is not as effective with modern refractory materials as with earthen finishes. See chapter 5 for repair methods.

(Many builders have built their manifold with masonry instead of using a second metal barrel. Sculpting a spacious manifold from cob is a special skill which is hard to convey in writing. It's easier to follow how to build a hollow brick plinth, either square, or roughly octagonal/round. An example brick manifold is shown in the photos above.)

Exit chimney and heat-exchange pipes should be laid and set to their proper slope. (See the diagram here). We work backwards from the through-roof or through-wall opening because the appropriate exit spot is

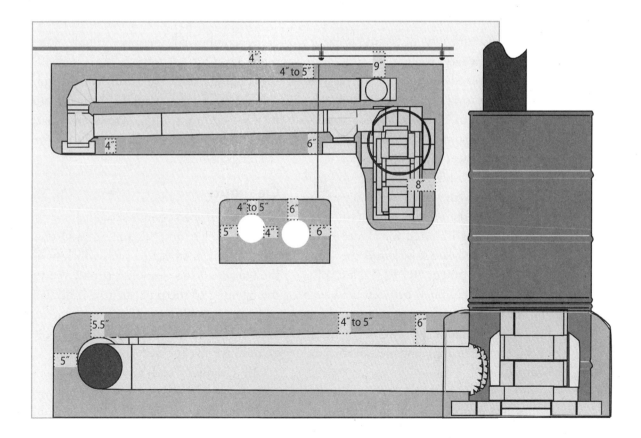

Materials Properties

Good conductors soak up and transfer heat. Good insulators do the opposite — they prevent conductive heat transfer.

The goal with mass heating is to conduct most of the heat into the mass for storage, while protecting critical differences in temperature with insulation.

Best Heat Storage to Best Insulation:

- Water, wet dirt (mud)
- Soapstone
- Granite, basalt, obsidian,
- Brick, clay
- Dense earthen masonry (thermal cob)
- Dense concrete
- Light earthen masonry (straw cob/adobe with straw), light concrete
- Pea gravel, crushed rock
- Dry dirt (loose soil), pumice, sand
- Insulation concrete (cement-perlite, cement-sawdust)
- Wood ash (loose)
- Vermiculite
- Rock wool (board or batt)
- Fiberglass, combustible fiber insulations (not for high-heat use)
- Perlite (clay stabilized)
- Perlite (loose)
- Refractory ceramic blanket/board

---

usually constrained by joists and rafters; it's easier to get a nice, straight, vertical chimney, with fewer elbows and cuts, if you work from the exit toward the unfinished bench.

Seal the heat-exchange pipes, working from the exit point back toward the manifold. Three sheet-metal screws per joint, and/or a layer of high-temperature foil tape will help keep the pipes from separating while you work. ("High temperature" in this case means it's rated for about 300°F, unless your hardware store stocks the "good stuff" rated for 500 or 600°F.) A telescoping section on the chimney, between the bench and the ceiling box, may be useful to create flexibility (for thermal flexing, frost heave, or building settling) and future maintenance access.

While sealing, check for combustible shims, and replace them with mortar bedding or noncombustible masonry rubble. Keep supports an inch or so away from seams for better airtight seals. We usually cut a custom connection to the manifold, being careful to minimize protruding screws and sharp edges that could snag someone while they are cleaning the heater.

Cleanouts

The cleanout access points should already be in place as T's; now it's time to add their caps. We prefer stovepipe caps, which fit on the outside of the cleanout opening (crimp the opening of the pipe slightly, if needed, to fit the cap). We avoid crimped ducting caps, as they do not seal well and can be difficult to remove for annual cleaning.

All cleanouts and any other working parts (such as the clamp for the removable barrel and lid) will need clearances around them for fingers or tools to access the parts. If desired, create removable formwork (from cardboard, plastic-lined rope, or rags) around these areas to hold the space during construction. (Wooden formwork can cause cracking in the masonry, and may create difficulties planning for its later removal.)

Pay special attention to barrel and manifold cleanouts — these are the most frequently used (for annual maintenance), yet it's critical to avoid air leaks in these areas. If the cap seems loose or the crimps on the duct are visible, you can install braided woodstove-door gasket inside the cleanout cap for a better airtight seal.

Thermal Mass
Surfaces

The surface of the bench may be brick, stonework, or earthen masonry to be plastered later. All masonry materials (including decorative finish materials such as tile, slate, or finish plasters) can be counted toward the final thermal mass thickness.

The original, most popular method for building comfortable and affordable benches is wet-formed earthen masonry, or "cob" (see Appendix 1). Cob is a clay-based concrete, mostly sand and gravel aggregates with about 10–20% clay. It is easy to shape around the pipes with minimal fuss and formwork, and eliminates the need to cut or dress masonry units. It holds heat very well — better than concrete, weight for weight — and is generally nontoxic and easy to use.

The only times we do not prefer cob is when the project is expected to remain damp throughout its working life (such as a greenhouse) or there are no suitable materials available locally — when a site has no access to mineral soils, or local soils are close to 100% silt. In these cases we may suggest using another type of masonry: either a casing of unit masonry (like stone or brick) with a tamped-earth infill around the ducts, or solid masonry of your choice.

With unit masonry casings (like brick or dressed stone), it's often easiest to build

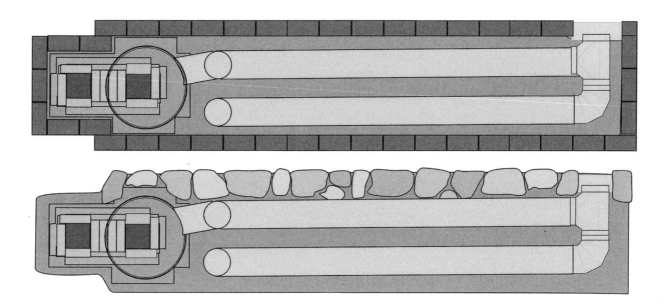

the casing first, with cleanouts in place, then fill around the pipes to bed them in cob or tamped earth.

Cob Method

- Mark out the intended finish or sub-finish dimensions on nearby walls (for areas where finish plasters or tile will be added later, allow ½" to 1" for finish plaster).
- Formwork can be used to set critical faces, like holding the back of a bench away from the wall for the 4" air gap. However, remember that earthen masonry does not set until moisture evaporates; damp-proof formwork will result in a project that dries very slowly.
- Lay any facing stone/brick that will be visible along the front edge.
- Keep cob out of contact with damp ground, and check slabs for damp (lay a sheet of plastic on the pad overnight, and check underneath for trapped moisture the following morning). When in doubt about the effective dryness of the footings (which is nearly always), we lay a first course of dry masonry to provide some drainage/ventilation under earthen materials. Puzzle-piece together any large pieces of rock, concrete rubble, brick, or other suitable materials to fill the space, using no mortar below or in between. Keep the outline tight, using mortar if needed around the outside edge, to discourage rodents or other pests from taking up residence. Fill any gaps in this dry footing course with rock, rubble, and large gravel, until there are no gaps larger than about ½".
- Spread cob/earthen mortar over the dry masonry base to make a bed for the ducting and next course. Build up 1" of fine earthen mortar around all ductwork, to double-seal and protect the ducts. (If any shims have a sharp edge against the pipes, replace them with mortar as you work.)
- Continue filling the core with masonry rubble and cob/mortar in courses. When time allows, we work about 6" courses and allow drying time in between. If time is short, we keep the cob quite stiff (like putty) with minimal moisture, to build 12" to 18" lifts per day.
- At the end of the day, leave the unfinished core rough, with holes or dimples to speed drying and create "keying" points to join the next layers. To add new material onto dry material, brush away any loose debris, and moisten the dry material with water or clay slip for good bonding. Keep all masonry materials slightly damp while working, for better bonding and less dust.
- Make sure to build up evenly on all sides of the duct to prevent punctures or squashing. Do not cover the top or fill between the ducts until the outer walls beside them are built up. We prefer to leave the pipes visible to prevent accidental crushing, until we have enough cob or stone ready to build a single layer 2" to 3" thick over the top of the ducts. The side walls and top act like an arch, supporting weight without crushing the duct. On a busy work site, we often put planks or plywood "spreaders" over the ducts once they are covered in wet masonry, to protect them from accidental damage until the cob hardens.
- The full, final thickness can be added now (5" to 6" total above the pipes), minus any

thickness reserved for finish plasters or tile work (usually about 1"). We often work from a level line along the wall, build up a perpendicular spine of level cob at the intended sub-finish height, and then fill in the surrounding areas to match that level. The sub-finish surface is usually left rough (allows faster drying and better attachment of finish plasters); within ¼" of level is good enough.
- Allow the core to dry completely (usually 1 to 2 weeks). The surface will dry sooner, but moisture continues to wick from inside. It is dry enough to finish when no more dark patches are visible, and the surface is no longer cool to the touch. Fans and test-firing the heater will speed drying.

Some moisture from combustion unit mortars and condensation usually finds its way out through the heat-exchange channels. "Drooling" usually stops once the core is completely dry and warm. We fire the heater up before completing the finish work to dry the core, set the pipes and barrel (in lieu of expansion joints), and make sure everything is working properly.

Firing-in the Whole Mass

Refractory vs. Earthen Masonry

Many modern refractory materials do not tolerate "wet-firing" — it can cause poor curing or steam spalling (small steam explosions that blast chunks of material loose from the surface). Earthen masonry is forgiving enough to tolerate wet-firing, and we use this method to allow for later thermal expansion.

If using modern refractories or other rigid masonry materials that require a longer cold cure before firing, or if you will not be able to fire the heater while the earthen material is still flexible, you may want to include expansion joints (as mentioned above during manifold construction). Critical areas include all edges where metal parts are buried in the masonry; and around the firebox where temperature differences are largest. Flexible insulation, braided fiberglass gasket, or other flexible and non-flammable materials laid along the edges and corners create the necessary "give" between dissimilar materials or temperatures.

Test-firing

For earthen masonry heaters, test-fire while the heater mass is still wet. This test with a cold wet mass (typically during summer, as it's the popular building season) will be the most difficult draft conditions the heater is ever likely to experience. A successful wet test predicts even better draft performance in winter conditions, with leeway for minor issues like storm winds or neglected maintenance.

Follow cold start procedures outlined in Chapter 5. Wet insulation material may temporarily cause excessive steam and smoke, or balky draft. Some refractory insulations are stabilized with rice starch; this burns out and makes black, sooty smoke in the exhaust over the first several firings (each time the system reaches a new high temperature, more soot burns away).

Performance Goals

A test-fire under the above conditions is successful when the barrel heats up quickly; the system drafts properly (fire is drawn down the burn tunnel toward the heat riser); and smoke clears (white or clear exhaust) within

3–5 minutes of lighting, except as noted for starchy insulation.

For cold-climate heaters, the exhaust at the exit chimney is about 100°F (may run 115–150°F once dry). Heaters that are used only occasionally, used in mild weather, or that may routinely be operated while damp would have shorter pipe runs and a higher exit temperature (see Chapter 6, "Channel Length, Chimney Design, and Temperature").

Failure Points

If the heater backflows, smokes, or does not draft well, try the following:

- Use cold start instructions from Chapter 5 — prime the chimney, etc.
- Check for blockages — in manifold area, firebox, and cleanouts; masonry may have dropped into the channels.
- Run the calculations from Chapter 6; check that the firebox proportions, channel length, and chimney height are correct.

If the heater overdrafts, that is, if it sucks up tinder and kindling so fast that paper won't stay in the feed area during wet test-firing on a warm day, draft may be excessive during cold weather. Consider lengthening the heat-exchange duct if possible. For further details, see Chapter 5's troubleshooting section and Chapter 6's rules of thumb.

Finishing

Barrel and Exposed Metal

Do not paint galvanized or sheet metal ducting to look like stovepipe — this creates a safety hazard for future owners and builders. Barrels and exposed pipe can be coated with high-temperature enamels (stove paints are available rated for 1200°F; engine-block paints rated for 2000°F are available in many colors).

For optimal radiant heat and nontoxic maintenance, we prefer a weathered-steel treatment. Clean the barrel thoroughly (see the "Pocket Rocket" sidebar earlier in this chapter), allow to dry, then coat with any cooking oil, like you would a cast-iron skillet. Some people like to use their well-cleaned barrel top as a griddle.

Masonry Finish Work

Casing

Please note that suitable finish materials must be compatible with surface temperatures (high around the combustion unit, lower in the bench area), and with the substrate (softer finish materials over earthen masonry, harder stuccos over cinderblock or cement).

Compatibility between earthen, Portland cement, and refractory masonry is discussed in Appendix 1. (There are known problems with mixing certain types of materials in certain situations.)

We do find that earthen materials are more comfortable for seating than cement or stone; the "give" is slight but makes a difference to tired joints over time. Wood trim and cushions can also be used where surface temperatures are appropriate.

Plaster Casing over Cob Core

Plasters can be made of the same clay-based minerals as cob, but sifted finer and with added fiber or binders according to the desired effects. Or they can be made with breathable natural materials such as lime or gypsum plasters. Mica, rice hulls, or other "glittery" materials can be included for visual texture if the finished plaster will be burnished and unpainted.

- Moisten the dry core with water or clay slip.
- Patch any cracks with dense cob (plenty of sandy aggregate), ripping out loose material if needed.
- We prefer to lay a fiber-rich scratch coat to toughen the seating surface, and key into the cob core. Apply a straw-rich cob with relatively small aggregate (¼"-minus or rough sand), or a rough earthen plaster with 1" or longer fiber.
- Allow this scratch coat to dry leather hard. Trim or fill any uneven parts.
- Add a finish layer of fine-sifted earthen plaster or lime plaster. If you have leftover clay-based plaster, especially if you have created a special tint or texture, make "biscuits" of dried plaster mix and store them with household paints and trim scraps for color-matching touch-ups.

Stone or Tile Facing Work

Earthen masonry can support thin tiles or mosaic work, but its bond is not strong enough to glue large slabs of facing stone in unstable orientations. If a stone finish is desired, it should be built up at the same time as the core of the mass, like a stone casement wall, with individual stones laid in stable orientation. We usually build stone or brick facades at the same time as the core, in courses.

Stone slab can also be placed on the top of the bench, as long as overhangs do not exceed about 3" (or 25% of the stone's width) of unsupported stone.

For facing stone laid at the same time as the bench, lay the outer casing in courses along with the earthen core. Work from the bottom up, with three main goals: stability; a nice flat facade; and finally, attractive distribution of your available stones.

Masonry made of natural rock requires some extra care to create stability. Here are some of the details you'll need to consider.

- **Three points:** Rest stones flat at the base, or fit them where they are cradled by three points of contact (no teetering). Mortar and wedges can help, but they will fall out more easily than large stones. If the stones are laid so they would be stable without mortar, the work will be more stable and durable. Mortar is used for mouse barriers, chinking, heat conduction, and as "liquid shims"; not as "rock glue."
- **Wide side down:** Lay stones on their widest side, so that they are wider than they are tall. Avoid standing rocks on edge.
- **Face work:** The bench face should be plumb (straight up and down), or the seat should hang out over the base a little. (Recessing the lower front of the bench slightly allows people to get their heels under them when standing up.) Some builders use a string line to check that rocks or bricks stay roughly in line.
- **Get it even:** Courses should be level, or slightly tilted toward the mass (you are building an arch around the pipes, so it's OK to lay the stones a little bit like a keystone arch, as long as the front of the bench stays in line). Avoid leaving a sloping surface like a mountain landslide; you won't be able to place the next stone without it sliding off.

- **Coursework:** Avoid running joins (vertical cracks that line up through several courses). Bridge gaps in a lower course by laying a stone in the next course across the gap. Use small stones or wedges as needed to match levels and support the next course.
- **Stretchers:** Bridge from the face work back into the main bench using long stones set with their short ends out. This braces the facade against tipping.
- **Faking it:** Where the available rocks won't do the job, you can cheat. Cut a rock to the shape you need, cement several rocks together, or sculpt a fake rock from plaster (later on). Some masons use epoxy to glue anchor points to flat rocks, so they can be set more deeply than the rock's natural shape would allow; just don't do this right near the hottest areas (firebox to manifold cleanout).
- **Think about critters:** Ensure there are no holes or crevices to attract pests. Make any intentional air channels large enough for a vacuum hose or other cleaning tools, or a hunting cat (about 4"). Removable screens or slatted vents may be used to exclude pests. Any small gaps should be too small for lost pencils or noisy vermin (<¼").
- **Showing off:** If placing a favorite rock before plastering, make sure it sticks out "proud" to match the final finish surface. Smaller, lightweight rocks (up to ½" thickness, or about 1 lb) can be set like tiles into the finish plaster, especially if they have rough backs or beveled edges that can grip the plaster.
- **Top slabs:** Work to a sub-finish layer; make sure there are no stones sticking up that would keep your top slabs or tile from setting flat. Lay a thick bed of mortar. If you like, use a cement trowel or rake to create ridges for easier leveling. Make sure the slabs are clean; some materials benefit from being damp before they are set. Set the slabs on the bedding, then tap down any proud corners to level the seat.
- **Keep it clean:** Wash stonework with a damp rag as you go along, or as soon after mortaring as practical. It may take several passes to completely remove mortars. Clay mortars can be cleaned at any point in the future, but tidy up immediately if using chemical-set mortars like lime, refractory cement, or Portland cement.
- **Mind the gaps:** You will inevitably end up with some gaps between the stones. A tinted finish plaster or breathable grout (lime, fine clay/sand/fiber) can be used to fill gaps and create consistent, color-matched joints. Pointing (the name for the decorative finish layer of mortar in brick or stonework) covers any rough spots and stabilizes the stones, like grout does for tile. Re-pointing may be needed occasionally if your stonework was less than stable, or if the stone face is exposed to weather, thumps, and bumps.
- **Make it shine?** A light coating of oil (linseed or cooking oil) can give an attractive, wet luster to facing stone, but should only be added after pointing and cleaning are complete. Waxes that are applied hot are not a good option, since they will melt again when the bench gets warm.

For Thin Tiles or Mosaic Work

Tiles of ¼" to ½" can be used with earthen finishes, bearing in mind that they will trap moisture and should not be added until the bench is completely dry. Take your time to find the right tiles and plan out their placement: sometimes, a strip of tile detailing is more attractive than a completely tiled surface.

- Complete the rough coat and leveling as for earthen plasters, above.
- Lay out the tiles or design you wish to use. A breathable, compatible backing (such as string or burlap) can be used, but avoid plastic backings if possible. Mark out placements, and remember that corners and edges may be affected by thickness of added material. Smaller tiles with more grout space are more breathable, and thus more compatible with earthen masonry. Tiles can also be set into plaster, with the plaster forming the main coating with a tile accent.
- Make a smooth finish plaster or mastic to use under and around the tiles. Do not use synthetic or vapor-barrier mastic over earthen masonry. Breathable mineral grouts include lime putty, finely-sifted earthen plaster, or a fine lime-sand grout. Synthetics and oil-based grouts should not be used, as they may not be heat-compatible. Grouts like these (or cement-based products) that trap moisture within the bench can lead to blistering and loosening of tiles.
- Prepare a breathable, wipe-on mineral grout for filling joints later if laying continuous tile or mosaic.
- Working in sections from one end to the other, wet the earthen masonry, and apply the plaster or undercoat with a concrete or tile trowel (jagged edges make a corduroy-like texture to aid in leveling tiles). Place the tiles as desired, waiting until a section is placed before tapping in any proud corners. Work any plaster areas between the tiles with a smooth float or tile.
- Once the tile areas have set firm, return to areas between adjacent tiles and fill the spaces with grout. Wipe off promptly with a wet finger, flexible silicone or rubber scraper, or damp rag.

Wood Trim

We have used wood trim successfully on several benches (see the Bonny 8″ and Cabin 8″ designs in Chapter 3 for examples). We have also helped other builders create some prototypes using wood boxes around the entire thermal mass (see Appendix 3). However, we have also seen charred or burning wood in some prototypes, especially near the intense heat of bell or firebox.

We recommend test-firing the almost-finished heater in realistic conditions before adding wood trim, and using wood trim only in areas that are known to maintain safe temperatures in a particular heater's actual operation.

Masonry thickness of 6″ (or more) should separate wood trim from heat-exchange channels. Wood trim should not be used in the clearance areas around bell and fire hearth.

Test-fire the heater several days in a row, until it is completely dry, checking that the sub-finish layers do not exceed 150°F throughout the firing cycle (including several hours after the fire). If the surface is too hot for wood, thicker masonry could help (add another inch or two of plaster and test again). Attractive and heat-tolerant alternatives to wood trim might include Saltillo tiles, marble or granite slabs in warm colors, "parquet" mosaic with streaked glass tiles, or even wood-textured concrete siding with heat-tolerant paint or stain.

Dead men (supports for wood) are laid in the rough or finish plaster layers.

- Prepare scrap wood with a bristling layer of bent nails, old screws, or natural branches short enough to embed and lock into the earthen plaster. Make sure the bristles are located and sized so they will not puncture any pipes or ductwork.
- Lay the wet plaster as above. Coat the roughened sides of the dead men with clay slip (or lime-water for lime plasters), and squoosh them in. Work the plaster around the spines, and tap to level.
- After the dead men are leveled and the material dries, wood trim is screwed or nailed into the dead men as for ordinary framing. (Pre-drill if needed, and use suitable-sized screws, to avoid strain that might bust the dead man loose from the masonry.)
- If wood trim or structural members are in contact or partially buried in cob (like posts for a nearby wall or roof), expose or wrap flexible expansion jointing such as insulation around at least one side, preferably two. This is to allow for different rates of thermal and moisture expansion. Driftwood mantelpieces and other organic shapes can easily be half-buried in cob.

Combustible supports should not be buried within a masonry heater, but wing walls (thin connecting walls) of cob may be used to extend the masonry into contact with existing walls.

Metal Trim and Fittings (Doors, Cleanout Covers, Etc.)

Metal fittings in high-heat areas can "crawl" over time due to thermal expansion.

The metal shapes we have buried in our cob so far are round, with no corners to start a fracture when they expand with heat. The buried parts are generally below 300°F, and there is at least 4" of cob around these structures to resist cracking.

If attaching metal details like a stove door or antique metal panels, plan for thermal expansion. Burying up to half of a metal object or leaving two faces exposed may allow it to expand without cracking the masonry. Flexible "cements" or welding are used for metal-to-metal seams, but there are limited materials for flexible masonry-to-metal

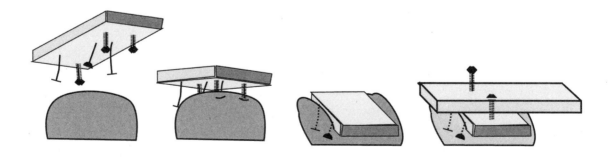

joints. Stove gasket or refractory felt may help offer expansion joints.

We test-fire our systems while the casing cob is "green" or leather hard, deliberately flexing the material to allow expansion without cracking. This technique is completely anathema to cement or lime-based masonry, as early heating interferes with the proper setting of the material.

To properly set a metal frame within masonry in a high-temperature area such as the feed lid:

Our finished example project, photo courtesy of Priscilla Smith.

- There must be a metal frame to receive the lid, anchored into the masonry deeply enough to allow for the force and movement without cracking masonry or pulling out anchors. Some builders create an independent cage of metal, cradling the firebox and surrounded in the casing masonry, padded with insulation blanket.
- Prepare the metal frame for heat expansion stress by attaching flexible expansion jointing (flexible fiberglass gasket, rock wool, or ceramic-fiber insulation) to any corners or edges that will be buried. If the material will probably expand in all directions, wrapping the entire part may be needed. If the material can be expected to expand linearly (like the lintel over a fireplace), pad the ends where they will meet the masonry. Also cover any hinges or moving parts with plastic or tape to exclude mortar.
- Prepare your mortar and masonry units. It is helpful if prior courses are firm and stable, but the metal part can be built in place along with any coursework at the same level. If the metal is supporting further coursework, allow the mortar and weight.
- Use fresh mortar, and dampen or roughen any cured masonry to ensure a good bond. You can paint the metal itself with a thin coating of mortar if desired. Seat the metal (with expansion joints in place) in the fresh mortar, gradually working it into place and checking level or plumb before tapping it into place. Make sure that no mortar gets on hinges or operable parts.
- To reduce cracking, use thicker masonry around any embedded corners, with well-placed, flexible expansion joint materials between the metal and masonry.

Other Decorations

Additional decorative touches may be added over time as you get used to the stove's performance, such as decorative screens or mosaic around the bell; cushions or futon pads; recoating or tinting the surfaces; etc. Cushions, fabric, and other removable decorations are covered in Chapter 5.

Heat riser: alternate A and B

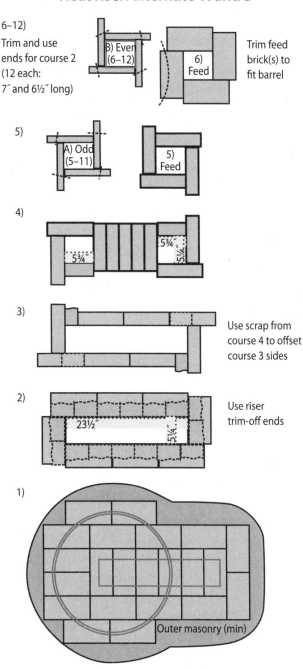

0) Non-combustible floor or foundation; Insulation (under and around entire firebox): either 2" of clay-stabilized perlite, or 1" refractory (e.g DuraBoard, rock wool).

6" heater in firebrick, top view of each course

6" Rocket Mass Heater

Sample firebrick layout for 6" combustion unit

Brick coursework:

Interior dimensions:

The burn channel is 5.25" × 5.75 inches: 5.75 inches tall (1.25" plus 4.5" plus mortar), 5.25" wide, and the openings at fuel feed and heat riser are 5.75" long.

The total burn tunnel floor length is 23.5 inches (a little more depending on mortar).

Fuel feed is 16" deep, the maximum in proportion to a heat riser 48" tall.

Note the alternating spiral layout allows leaving the ends sticking out; any dimension of brick can be used. Only 2 cuts are required (5.75" bricks that butt against the bridge, course 4). But if you trim the heat riser bricks, their ends can be used in course 2, saving a few bricks and improving the insulation fit and manifold clearances.

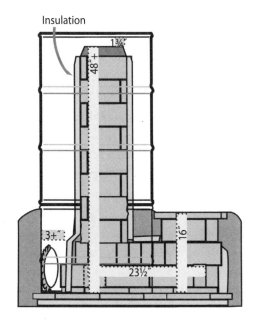

6" heater in firebrick, side view

Step-by-Step Construction Example

97

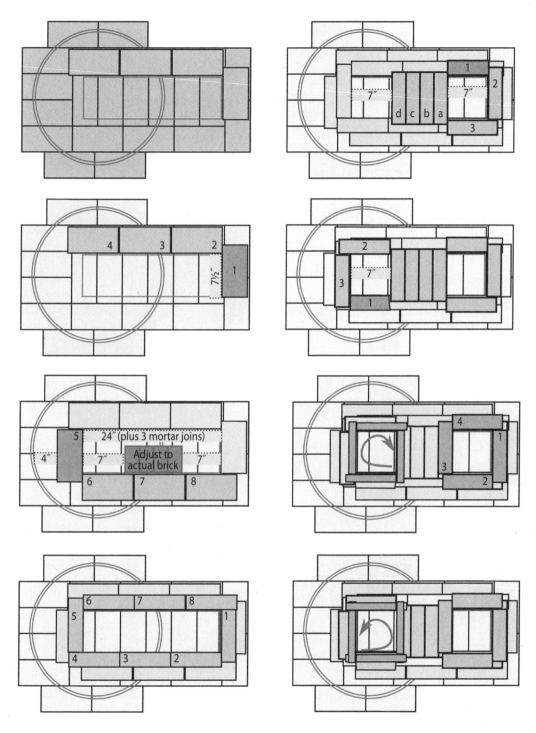

8" firebox, one-page cheat sheet for brick layout. Read from top-left down to bottom, then top-right down to end. (The bottom-right final course could be useful, enlarged, as a scale template for project layout.)

6" heater layout for firebrick and cut-barrel manifold.

6" (150 mm) Rocker Mass Heater

Sample combustion unit layout

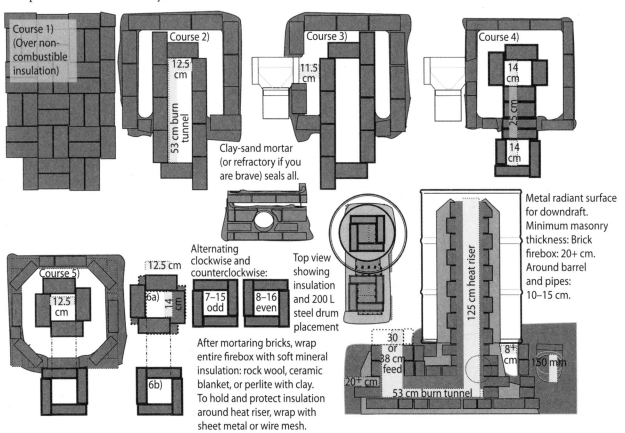

A 6-inch combustion unit in red brick, with a square brick manifold. Inch dimensions: Burn tunnel 21" long by 5" wide, minimum 3" gap to pipe side of manifold. Heat riser and feed interior: 5" wide by 5.5" long. Heat riser at least 47" tall, feed 12" or 15" tall.

8" Rocket Mass Heater

Sample brick layout for combustion unit

Courses:
Heat riser continues alternating A and B.

Brick coursework:

Interior dimensions:

The burn channel is 7" × 7.5 inches:

7+ inches tall (2.5" plus 4.5" plus mortar), 7.5" wide, and the openings at fuel feed and heat riser are 7" long.

The total burn tunnel floor length is 24 inches.

Fuel feed is 16" deep, the maximum in proportion to a heat riser 48" tall.

Note the alternating spiral layout allows any dimension of brick to be used, with only 2 cuts required (7" bricks that butt against the bridge, course 4). Other bricks' ends or corners may be trimmed for tighter clearances if desired.

Firebrick can be laid with fireclay slip, or with a thin clay-sand or fireclay mortar for bonding, seal, and stability. Around the brick, the insulation also serves as a secondary seal and expansion joint.

Check often for flush interior walls, and level courses.

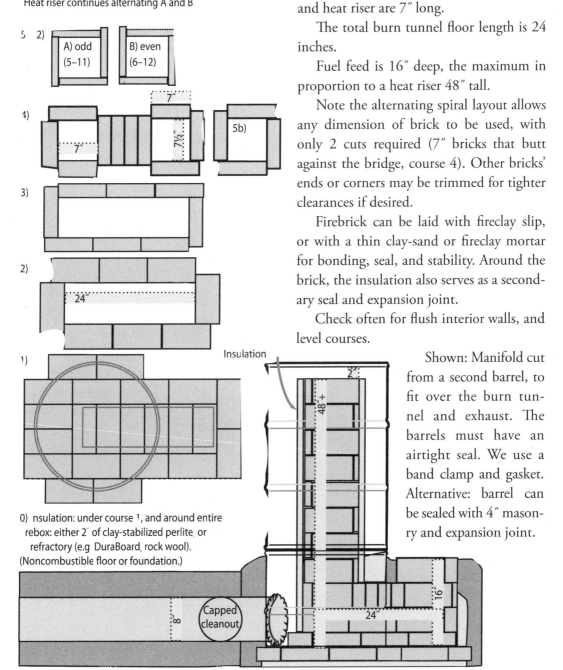

Shown: Manifold cut from a second barrel, to fit over the burn tunnel and exhaust. The barrels must have an airtight seal. We use a band clamp and gasket. Alternative: barrel can be sealed with 4" masonry and expansion joint.

8" Rocket Mass Heater (200 mm)

These dimensions wre originally developed in inches (8" or 6" diameter stovepipe, 9" by 4.5" by 2.5" firebrick). Luckily, metric equivalent parts are similar and common.

These firebox dimensions are for a rocket mass heater using 200 mm ID (internal dimension/diameter) stovepipe for heat exchange channels and exhaust chimney.

The layouts can be built with almost any dimension of brick.

Mortar thickness or brick dimensions can affect the buildable dimensions in courses 2, 3, and 4.

Before starting, lay together the 4 bricks on edge for the bridge (course 4), with mortar if used. Measure them. Is the span 24 cm as shown? These bricks plus 18 com* openings make up the total firebox length. With mortar, the burn tunnel might be 61.5 cm long instead of 60. Work to the actual length in courses 2 and 3.

Two bricks must be cut to make the short (18 cm*) sides of the firebox openings at course 4.

In courses 2 and 3, the total height of the bricks may require adjustment to the width and opening length. Length of openings = burn tunnel height (shown at 18 cm). Multiply either by the 19 cm width = flow area. If course 2 and 3 are not exactly 18 cm tall, find the closest stackable dimension, and adjust opening length to match. Adjust width to compensate, to keep the flow area roughly 330 cm^2.

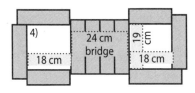

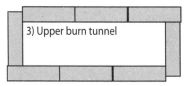

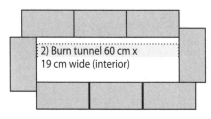

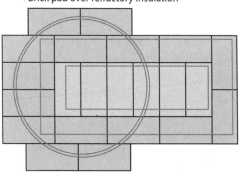

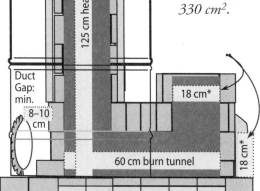

Insulation shown: Slices of insulating kiln brick. The whole rests on a non-combustible footing e.g. concrete slab, with 1" to 2" of stabilized perlite for leveling, or 1" refractory insulation board.
Adapted from Greenhouse 8" Rocket Mass Heater © 2015 E&E Wisner http://www.ErnieAndErica.info

Chapter 5

Operation and Maintenance

THIS CHAPTER IS INTENDED TO SERVE as a stand-alone guide for owners of rocket mass heaters. It mainly refers to J-style rocket mass heaters built in accordance with guidelines in this book.

Be aware that this chapter may contain information about features or methods that are not relevant to your particular heater. Heaters with unconventional chimneys, odd proportions, or other improvised alterations may not operate as described (or at all).

For general troubleshooting tips for *non-standard* heaters, we recommend the FAQ section of *Rocket Mass Heaters* by Evans and Jackson. You can find discussions about improvised heaters and stoves online at energy forums like www.permies.com.

In order to properly maintain your heater, you will need to document its construction details for future reference. Don't rely on your memory. As a builder/owner, you should consider it your responsibility to document your project's site-specific information, which is essential to create an accurate, unique manual for your particular heater. Someone other than you might need this information in the future. In some instances, we've provided blank fill-in lines to prompt documentation, but you should also attach your own project plans.

In this chapter, we will cover general operating and maintenance issues, and help you create an owner's manual for your specific heater.

Warnings

Heating involves well-known hazards; these are the operator's responsibility to manage. The laws of nature are not subject to lawsuit or appeal. Remember:

- **Fire is dangerous.** By lighting a fire anywhere, you become responsible for tending it. Check and remove nearby combustibles before lighting any fire. Watch it; feed it only appropriate fuels; and make sure it's finally out. Never leave fire unattended.
- **Smoke is nasty.** Smoke carries heat, toxic and flammable vapors, carcinogens, heavy

metals, depleted oxygen, invisible carbon monoxide (CO) and CO_2. The operator must ensure proper ventilation and eliminate all preventable smoke — not just indoors, but residual smoke that can coat channels with creosote or pollute shared outdoor air.

- **Metal and masonry get hot.** Keep unsuitable items away from hot surfaces. Observe the actual operating temperatures of your particular heater — during and after firing. Monitor for changes if fuel, weather, or operating patterns change.
- **Wood burns.** Natural fuels are less predictable than refined fuels; the operator is responsible for fuel choice, dry storage, and fire tending. Store fuel and other flammable materials away from heat and sparks.
- **Dust chokes.** Avoid breathing anything other than air. Avoid smoke, masonry dust, ash, insulation fibers, and funky upholstery. Use water to control dust during masonry work, and use respiratory protection and good vacuum filters during maintenance.
- **Water erodes.** Watch for signs of water damage, leaks, or corrosion — especially in areas where your stovepipe goes through the walls or roof, exposed joints, cleanouts, and any areas where spills, condensation, or rising damp could collect and pool.
- **People play.** Enjoy your heater. Engage with it. Anticipate the natural curiosity of visitors (especially children) who may play with the firebox, fiddle with cleanouts or dampers, or drop toys in odd places. Check the heater carefully after guests visit.

These are radiant heaters, not automated furnaces or boilers. They rely on manual operation for safe and effective performance. Although they are designed by and for observant, responsible operators, for many people the operating principles are counter-intuitive at first. Operation becomes easier with knowledge and training, but more importantly, with practice. An attitude of cheerful problem-solving goes a long way.

Other owners, builders, and designers will probably be happy to offer advice if you encounter any difficulties. Please share the benefits of your own experience the same way. We're all in it together; the risks and benefits involved affect everyone.

Do your best, ask questions, and compare notes, and then decide for yourself whether this is the right heater for your situation.

Be safe. Have fun.

This guide is intended to assist, not replace, your own observation, skills, and experience. Every site-built heater is unique. Please write your own notes, and adapt this guide as needed. Non-commercial sharing and adaptation is permitted and encouraged. However, because every rocket mass heater is, by definition, unique, the authors can take no responsibility for the accuracy, safety, or applicability of this information to a particular situation. No persons associated with this publication can be held liable for injury, damages, or deaths resulting from this information or related activities.

Types of Heaters

The laws of physics dictate how rocket mass heaters work. In order to operate one successfully, it helps to understand that there are different types of heaters, and that the design decisions behind rocket mass heaters are made for specific reasons.

Some Useful Definitions

A *heater* is designed to heat space or things. Common space heaters include fireplaces, woodstoves, electric or gas radiators, baseboard heaters, radiant floors, etc. Personal heaters include heated seats or saunas, electric blankets or heating pads, or hot water bottles.

A *stove* produces quick heat for cooking or heating. *Stovetop cooking* usually involves heating from underneath, as contrasted with baking (in the oven). Many stoves are transportable, but large site-built masonry heaters may also be called *masonry stoves*.

An *oven* is a warm or hot enclosure for baking or drying material.

A *furnace* blows warm air through ductwork to heat a building. Furnaces are usually located in a space separated from the occupied spaces of the home (in a closet, utility room, or basement). (Industrial furnaces are different; they can be immense, usually custom-made facilities for smelting, refining, and other high-heat processes.)

A *boiler* can also be used to heat buildings, and it also operates separated from occupied spaces (often in a separate building); but instead of hot air, it produces hot fluids such as water or steam for distribution to the areas needing heat. (Even if the heated water never boils, this type of heater can still be called a "boiler.")

Rocket mass heaters are designed to bridge the gap between the nature of fire and human needs. Their design makes a few assumptions:

- **Steady warmth is more comfortable than alternating hot and cold.** Thermal mass stabilizes temperatures over time, saving a lot of fuel in seasons with variable weather. If you don't need steady heat, just a few hours here and there, then this design may not be your best option. See "Cold Starts" later on in this chapter for some methods for working around the cold mass, or you may prefer a lower-mass option like a parlor stove or radiant fireplace for quick, on-demand heat.

- **The goal is to heat people, not space.** Heat stored in masonry is of little value if nobody is there to enjoy it. Most people seek indoor warmth at predictable times: chilly evenings and nights, while resting after meals, and during sedentary work or relaxation. A heated bench or bed delivers heat where it is most appreciated, and in a place where it's generally convenient to tend to. Heaters located in a basement, crawl space, or separate addition may squander as much as three quarters of their heat output before any trace of warmth reaches the occupants.

- **Automation is not always desirable.** While automation is a convenient way to use fossil fuels, it does cost energy, and it also involves moving parts that can fail (or stop working in power outages). The rocket mass heater is designed to operate on natural convection draft with the help of a responsible operator — no outside machinery required.

- **Change is stressful, and your own experience will dictate how big a change this is.** Operators accustomed to heating with a conventional, low-mass woodstove generally find that rocket mass heaters offer far more comfort with less effort. Owners accustomed to automated central heating may find that heating with wood is more work or hassle than they expected; you earn the cost savings with some manual labor. Some owners use their rocket mass heater when convenient, and keep their less-efficient central heating set on low for backup. If you are happy with your existing heater, woodstove, or furnace, we generally suggest keeping it while you become accustomed to the rocket mass heater.

- **Elegant design (the simplest effective solution) comes from being in harmony with natural laws and human nature.** Conventional heating, ventilating, and air conditioning systems (HVAC) use abundant fossil fuels to overcome uncomfortable weather or poor building design. Ironically, this fuel consumption and the resulting greenhouse gas emissions may be contributing to more extreme weather events.

Rocket Mass Heaters

Rocket Mass Heaters are highly efficient solid-fueled heaters. They are designed to warm people, plants, and buildings with constant, steady heat, while requiring minimal fuel, pollution, or operator effort.

Rocket mass heaters have two main components: a combustion unit for clean burning of fuel; and a heat-exchange mass to capture and store that heat for gradual release. They will each have site-specific layouts, features, and proportions.

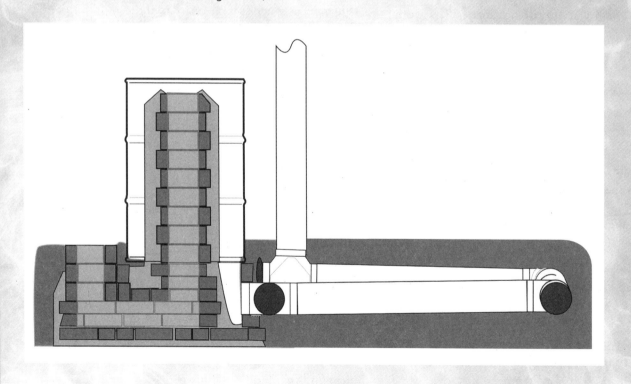

Rocket mass heaters use minimal resources to extract heat from abundant, renewable natural fuels, then convert most of that heat into efficient, ergonomic comfort. (Solar energy and ambient warmth can also be stored in the heater mass; see Chapters 2 and 4 for ideas about passive solar design.)

The rocket mass heater is designed around human needs, wants, and physical abilities that are common, but not universal. We hope that you'll consider whether it's the right tool for your situation, and if not, that these points may help you to find the right solution for you.

Rocket Mass Heater Components
Terms and Definitions
Combustion Unit: Fuel loads vertically in the *fuel feed,* burning hottest at the bottom. Flames move sideways along the *burn tunnel.* The insulated *heat riser* draws flames up to complete a clean burn. The *barrel* or *bell* radiates heat outward, and downdrafts the exhaust gases to the *manifold.*

Heat Exchanger: From the manifold, hot exhaust gases pass into the heat-exchange ducting, slowly losing heat to the surrounding thermal mass before reaching the exit chimney. The *heat-exchange mass* stores heat for contact comfort and long-term warmth. It may be a heated bench, floor, or bed.

Cleanouts: These are openings in the masonry that give access to the heat-exchange channels, one access point at each turn and each vertical drop. Some heaters may have an *ash trap,* a *bypass damper,* or *outside air intakes.*

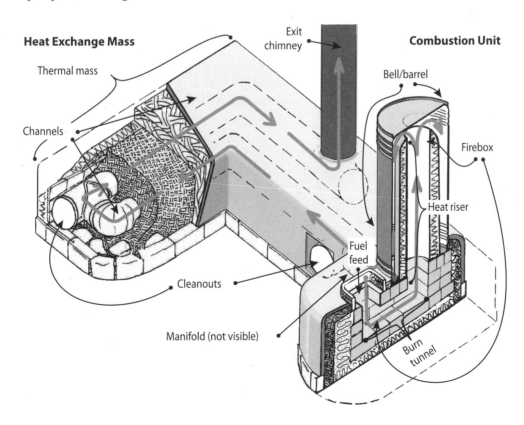

System Size: The inch measurement in a "6" system" or "8" system" refers to the interior diameter of pipe used for the exhaust channels and exit chimney. The cross-sectional area (CSA) of this pipe represents the desired flow area for all channels in the system, including the combustion channels. An existing chimney flue, or your desired heat output, will determine your choice of system size.

- 6" system: CSA about 30 square inches (28)
- 7" system: CSA about 40 square inches (38)
- 8" system: CSA about 50 square inches (52)

As-built Drawings

As-built drawings show what actually got built (not just our plans or good intentions). In case of later troubleshooting or renovations, accurate notes about what's inside that mass of masonry may save tons of unnecessary work.

Architects and engineers are trained to produce accurate as-built drawings. As an owner or builder, don't worry about artistry, just take good, practical notes. You can write the measurements on a pencil sketch, or take photos at each stage of the project. Put a ruler or coin in the photo for scale. If necessary, stick a tape measure down the tubes after the fact, and capture any missing numbers.

Show these details:

- Top view (plan)
- Side view (section or elevation)
- Materials and construction details
- Dimensions of the heater, thermal mass, and surroundings
- Combustion unit dimensions
- Cleanout locations
- Foundations
- Exit chimney or vent
- Any special features such as:
    - bypass damper
    - heat shielding
    - outside air, room air, fans
    - hidden access panels
- Any hidden hazards (utilities, asbestos, glass, or sharp infill)

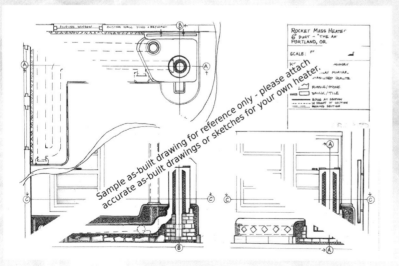

Sample as-built drawing for reference only - please attach accurate as-built drawings or sketches for your own heater.

Operation and Maintenance

Systems as small as 4″ and as large as 12″ have been built; but these present unique problems outside the scope of this book.

Heat-Exchange Ducting: Most rocket mass heaters have 20–40 feet of lined channel snaking through the heat-exchange mass. The length, material, turns, and cross-sectional area of this channel all affect the heat storage and draft performance.

(Note: As you build, be sure to note for your records any special features like bypass damper, branching channels, or large cavities.)

Thermal Storage Mass: Most heaters have 30 to 100 cubic feet of masonry mass, or 1 to 5 tons, with at least 4″ of masonry wall around the outside of each heat-exchange duct for structural integrity. (Low-clearance heaters generally have 5″ of masonry thickness plus a 4″ air gap to combustibles.)

Combustion Area Proportions: Correct proportions are critical to performance.

Heater Summary

Heat-exchange mass

Duct size: _____

Duct length: _____

Thermal storage material, size, weight: _____
_____

Combustion unit

Heat riser height: _____

Fuel feed height: _____

Burn tunnel length: _____

Firebox opening size:

width: _____ length/height*: _____

feed length and tunnel height typically match; note if different

Gap above heat riser: _____

Gap in manifold (minimum clearance): _____

Site Details

Project dates: _____

Location: _____

Building size: _____

Chimney height: _____

Foundations: _____

Attach as-built drawings or photos, or indicate where to find them.

Heater History

Attach additional pages as needed.

Initial Documentation

List names and contact info for people involved with the heater's installation and anyone who might be involved in repairs:

Designer: _____

Book, plans, or other design references: _____

_____

Builder(s): _____

_____

Helpers: _____

_____

Were there any contracts, guarantees, or warranties? Technical support?

_____

_____

Construction Notes

Where do we keep records of the design and building process? (Attach them, if possible.)

_____

_____

_____

Was the heater officially permitted, approved, or exempted? _____

_____

Was the design changed to meet local codes? Y / N Did these changes affect performance? How?

_____

_____

Are there any potential hazards, hidden features, or warnings? (e.g. electrical wiring, hidden dampers or cleanouts, buried glass or other hazards): _____

_____

Operator Notes

Tinder

How many sheets of newspaper/paper bags does it take to start this heater? _____

Kindling

Does it like a lot of kindling, or a few scraps? (at first _____; once warm, _____)

Fuel loading

Does this heater perform best (as most do) with a full fuel load? Y/N _____% full with _____ type(s) of wood

Fuel types

Any favorite fuel blends, types, or sizes?

Twigs/sticks/poles/log(s)? _____

Hardwood/softwood/mixed? _____

Bark-on/split/scrap lumber/paper? _____

Air controls

Performs well full-open? Y / N

What air setting makes no smoke, indoors or out? ←

Weather

Does weather affect this heater's performance?

Wind force/direction _____

Outdoor temperature _____

Rain or snow _____

(Note: If you notice changes in performance, or have to change fuel, air, or other control settings without any clear reason, this may indicate it's cleaning time or other maintenance is needed.)

Cleaning

What part(s) need to be cleaned out most often?

Firebox _____ Manifold _____ Cleanouts

What depth of ash do we leave in the firebox? _____

Troubleshooting

Does the heater have any special tricks for handling draft, smokeback, or cold starts? _____

Misc. observations

_____
_____
_____
_____
_____
_____

Documenting Your Site-built Heater

As the owner or builder, you need to document measurements and the materials and methods used in constructing your heater. It will be crucial to future maintenance. You'll want to have a file where you will keep all your plans, as-built drawings, project photos, etc. Use the "Heater Summary" shown here to get started on your documentation.

Combustion Area Proportions: Document heights, lengths, and gaps in the heater summary.

Operation

Most owners run their heater while relaxing at home in the evening, or at any convenient time of day during indoor office work, reading, or entertainment.

Rocket mass heaters generally:

- Are fired intermittently: full heat for a few hours, then close down between firings
- Stay warm for many hours (or days) after each firing
- Take time to "recharge" with heat if they have been allowed to cool completely
- Do not need to burn at night
- Should not be left burning unattended

For best results you should inspect the heater before lighting (see inspection list, below). Always operate with about ¼" of ash in firebox (a little ash insulates the fire and bricks, but anything more than 1" can choke the fire; so, remove excess ash before starting a new fire).

Routine Operation: Lighting the Fire

(See also color insert.)

- Use the inspection checklist (page 110) to make sure all the parts of the heater are in working order.

Preliminary Inspection

Look over the entire heater, with attention to the following:

All surfaces

_____ no cracks or damage

_____ no evidence of leaks (soot or smoke streaks outside, cleared streaks in soot inside)

_____ no heat-sensitive items (cloth, wax, paper) in wrong places

Cleanouts

Mark locations on heater drawings (C1, C2, . . .).

_____ Use cleanout openings to confirm that system is clean and free of debris.

_____ Close cleanout caps securely before lighting the heater.

Soot and fly ash normally accumulate between annual cleanings. Streak or spray patterns in soot indicate air leaks, which must be sealed. Sticky creosote should never occur; it may indicate problems with heater, or damp/green fuels. Outdoor debris (leaves, nests) may indicate a missing exhaust screen.

Fuel feed and combustion area

_____ Fuel feed lid or bricks to cover opening, adjust incoming air.

_____ Fireplace tools present and in good condition:

_____ long-handled tongs, _____ metal ash bucket (optional: _____ hatchet, _____ hearth brush, _____ ash scoop or shovel)

_____ Tinder and fuel: Fuel must be dry, sized to fit in the fuel feed, and stored a safe distance from the heater.

_____ Clear any combustible items from fuel feed/hearth area (4″ air gap around masonry, 18″ hearth around fuel feed).

Radiant metal barrel/bell

_____ Clear any heat-sensitive items from top and sides. Clear any combustibles within 3 feet of metal radiant surfaces. *(Do not dry fuel on, or in contact with, radiant metal surfaces.)*

Optional features (not all heaters have these)

_____ Wood trim: Good condition, no discoloration or warping near heat. *(All wood should stay cooler than 150°F (65°C) throughout heating cycle.)*

_____ Upholstery: Good condition and properly placed on the zero-clearance bench area (not against the barrel or fuel feed hearth). *(Use heat-tolerant fabrics; see "Upholstery" below.)*

_____ Fresh air channels including outside air, room air, heat shield air gaps. *(Keep all air gaps clear of loose items and flammable debris such as lint, dust bunnies.)*

_____ Heat shields: minimum 1″ air gap behind, below, and above for convection.

_____ Bypass damper: Use nearby cleanout(s) to inspect damper condition, if possible. *(Set to correct position (open for cold start, close to store heat.)*

 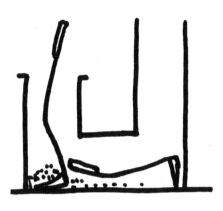

1. Inspect the heater and clean the firebox (leave ¼" of ash).

- Check for warmth. Most heaters start easiest when the mass and heat riser are already warm. If any parts of the heater are unusually hot, another fire could create an uncomfortable amount of heat. If the heater is cooler than the outside air, it may require extra priming to overcome this "cold start." Several methods for cold starts are described later on.
- Check the draft. Light a candle or twist of newspaper, and tuck it down through the feed opening into the burn tunnel to see if the draft is working. When the flame flickers away from you down the burn tunnel, the draft is working properly.
- Prepare your tinder and light the fire. Use only dry, natural firewood that fits in the firebox. Oversized, wet, or contaminated fuels can create dangerous problems.
- Place wood vertically, with thick ends down (to allow it to drop instead of wedging). Keep the actively burning wood in front (toward the burn tunnel and bell). As the kindling catches, stand a few small sticks behind the kindling, then add a few larger sticks behind that to fill the feed tube.

2. Use a candle or tinder to establish draft. Tuck lighted tinder partway into the burn tunnel. Draft is working when the flames pull away from you, toward the heat riser.

3. Drop kindling onto the burning tinder. Set the sticks vertically against the tunnel opening.

4. Once kindling is burning, add larger fuel behind the fire.

5. Load fresh logs behind the burning fuel. Reload as needed, using the largest practical logs.

- Remain near the fire for the first phase of the burn, about 20 minutes, to ensure that all kindling catches and drops down properly. Rattle the kindling if needed, to help it fall downward.
- Always add fresh wood behind the existing fire; don't disrupt the hot flame with cold material. If needed, move the burning logs forward with tongs so you can load the next piece behind them. Once the fire is well established, use larger pieces of wood (2″ to 5″ thick).
- Eliminate any vertical "chimneys" between pieces of wood. Wood that is too tall for the firebox, or unusually pitchy, may draw flames upward. Wood with a wet core may also refuse to drop and cause problems. Most problem wood can still be burned, just cut it down to a shorter length (ideally from the beginning; shorter wood dries quicker, as well).
- Most rocket mass heaters burn best with a full fuel load and an air opening between full-open and about 25% open. If necessary, use a brick or tile to partially cover the opening. We generally leave it at least one third open throughout the burn.

- Once the house is comfortable and enough heat is stored, stop feeding the fire. Reduce the air as the fuel dwindles to maintain a clean, balanced burn.
- When the fire is completely out, shut the feed lid to stop the air flow.

Emergency shut-downs are not recommended. If you suddenly have to leave before the fire is finished, you can close the opening down to a screened air slot (1" or so) and let the fire burn itself out. Do not smother an active fire by sealing off its air supply; this is likely to cause smoke and creosote problems in the stove, and possibly smoke leaks into the room if the chimneys cool too quickly.

If there is a minor problem, such as slow draft that seems to call for a maintenance inspection, the general procedure is to ventilate the room well to remove any smoke, close down the air supply to about 20% of the opening.

In case of a dire emergency, where immediate shut-down is critical, some operators have successfully used a metal bucket and tongs to collect burning fuels and remove them quickly to outdoors, and then extinguish them outside. You could also use a dry fog or powder extinguisher if needed (ABC type). In either case, ventilate the residual smoke from the room.

We do not recommend using water as an extinguisher for indoor fire places or stoves of any kind, as it creates a cloud of noxious steam and smoke that is usually worse than anything the fire was doing on its own. Water and steam can permanently damage hot masonry or metal.

After the firebox is hot, it turns kindling into coals very quickly. Large amounts of kindling, wood chip, junk mail, or other small fuels can create a blockage of ash and embers. Use a poker to clear an airway, and allow the fire to burn down before adding any more fuel.

To avoid this problem, load larger logs instead of small fuels once the firebox is hot.

6. Adjust air flow if needed — at least ⅓ open throughout the main burn. A full fuel load and full-open air allows the fire to self-regulate.

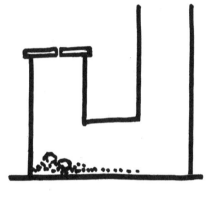

7. As the fire burns down, reduce air flow.

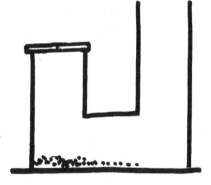

8. When the fire is done, close the feed opening.

And to reduce ash buildup in all parts of the system, use the cleanest tinder you can get. Avoid color-printed material (contains clay and weird minerals). Plain newsprint, brown paper bags, or shredded natural fiber make excellent tinder with little ash. Small quantities of butter wrappers, grease paper, or sawdust/wax firelighters are helpful too.

Hot Spots

- Barrel (200–800°F/90–400°C)
- Feed opening (radiant heat from coals that are up to 1800°F/1000°C)
- Combustion area masonry, cleanouts (150–200°F/ 65–90°C)
- Hidden voids or insulated areas, such as under upholstery

Trouble Spots

- Random items left on top of the barrel, or leaning against it
- Toys, flammable tools, or fuel scraps around feed area
- Fuel, upholstery, clothing, or other flammable items too close to hot spots
- Cleanout covers missing or loose

Spots to check.

Cold Starts and Optional Priming Methods

Cold Starts

A "cold start" is what you have when you are trying to light the heater when the mass of the heater is colder than outside air. Cold starts affect fireplaces and woodstoves too. A cold chimney will draw cool air downwards, instead of warm air upwards. It can be difficult to reverse that flow when you need to start the proper draft for your fire.

Cold starts most commonly happen under specific circumstances:

- during the first fire of the season
- when the mass is damp (as in a new heater being built, or a greenhouse bench under watered plants)
- in occasional-use heaters such as vacation cabins
- when the heater is fired during warmer weather, for whatever reason (mild climates, sporadic heating needs, or show-and-tell)

Unlike most fireplaces and woodstoves, the rocket mass heater has an "airlock" profile (like the U-trap in a kitchen sink). It drafts beautifully when hot, but slowly when cold. This helps the fire self-regulate its air intake, and prevents the heater from removing warm air from your home if you forget to close the feed lid after a fire. But the airlock feature does make cold starts more challenging.

Cold air can act as a dense "plug," pooling and swirling like fog in a valley. Trying to force warm air through a cold plug can produce some very weird effects, like smoke "bouncing" out of the fuel feed after a fire had just started to draw.

The secret to cold starts is to prime the vertical chimney(s) slowly, preheating the heat riser and exit chimney to draw cold air out of the system before lighting a larger fire.

Cold Start Methods

Paper Pusher: Classic Fireplace Method

Light a twist of newspaper under the heat riser as you would in an ordinary fireplace throat, to confirm good draft before starting the fire. When the flames draw properly (away from you, toward the heat riser), then add kindling and light the fire.

Candle Method: Simple and Sweet

Light a candle under the heat riser. Wait about 10 minutes, then pull it out and check. When the flame points toward the heat riser instead of back at you, the system is primed and ready. This may take up to 40

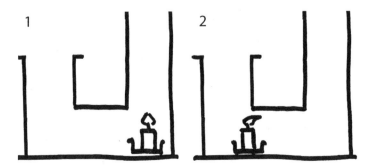

minutes, but we've never needed a second candle.

Chimney-priming Method: Rough-and-ready

Remove the cleanout cap at the bottom of the exit chimney, and light a small fire right inside the chimney. Burn a wad of newspaper, a Sterno stove, or some other small, clean fire right in the exit chimney. If possible, replace the cap while this little fire is still burning, to forcibly draw the cold air out of the thermal mass and replace it with warmer room air.

Chimney-priming fires work well in combination with the candle method. Be aware that if you try to light a priming fire in the chimney after a failed full-size fire the spent exhaust may put out the priming fire.

are warm, gradually close the damper to force warm exhaust into the heat-exchange channels. It may take some time to fully circulate the cold, stale air out and establish normal draft.

Bypass dampers usually follow the general valve convention: when the handle is lined up with the pipe, the damper is open for flow in that direction.

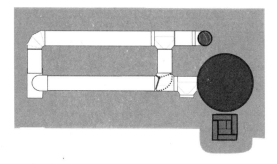

Horizontal Dump: A Workaround

A few systems have a horizontal exhaust exit that opens outdoors, such as a cleanout at the bottom of the exit chimney. On a calm day, it may be easier to start the heater with this cleanout open, "dumping" the cold exhaust (1) until the thermal mass warms.

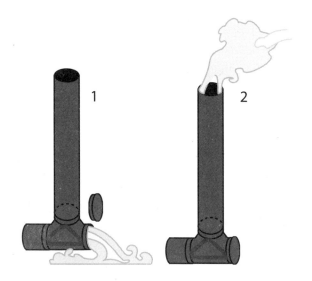

Bypass: Planning Ahead

Some heaters are built with a bypass: a shortcut that allows the exhaust directly into the chimney, bypassing the longer heat-exchange channels.

With the bypass damper open (as shown), the system can get going very quickly. Once the combustion unit and chimney

Once the exhaust coming out feels nice and warm, and rises as it floats away, gradually replace the cap and watch the exhaust rise up the chimney.

Weather Watch Method: Adapt to Actual

For the first fire of the season, wait until the cool of the evening, or start the fire early in the morning. Since the mass reverts to the average ambient temperature, not the coldest temperature, there is usually a time of day when the outside air is colder than the indoor mass.

Passive solar building design can allow the heater to soak up sun rays during winter days, stacking the deck so you never have a cold start.

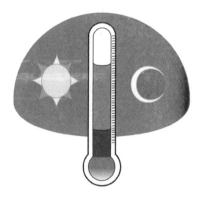

Power Assist Methods: The Electric Crutch

Normally we don't advocate electronic gadgets on wood-fired heaters. (What good is wood heat if it doesn't work in a power failure? Also, moving parts such as fans are often the first things to break down in the hot, corrosive atmosphere of combustion exhaust.)

Exceptions might include a cold-start booster fan that is only used occasionally (with battery or hand-crank backup), or an electrical preheater or chimney warmer (as simple as an incandescent light bulb or heat lamp left in place for a while and then removed before lighting the fire).

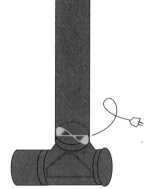

Design Features

The following will make heaters more likely to draft strongly in all conditions, including cold starts:

- a self-priming exit chimney next to the warm barrel (with or without a bypass)
- shorter heat-exchange ducting
- taller and better-insulated exit chimneys and heat risers

Fuel Selection and Storage

All fuel must be dry — below 15% moisture content, the drier the better. To check your firewood's moisture content, split a log and weigh the pieces. Then dry them on "low" in a bake-oven for a few days, or indoors with a fan for about 2 weeks. Weigh them again. The lost weight (water) should be less than 10% of the total (some water may remain trapped in the wood, depending on indoor humidity).

Season natural firewood in dry storage for at least six months prior to burning, up to two years if possible. Firewood stored in contact with the ground or wrapped in plastic does not dry or season properly.

Good wood storage provides excellent ventilation below and beside the wood, protection from rain and snow, and space to separate last year's dry wood from this year's green harvest. An ideal woodshed also includes tool storage and room to swing an ax in the shade.

Recommended Fuels

You can use any dry, natural wood fuel that fits in your fuel feed (taller pieces are a fire hazard). Hardwoods, softwoods, fruitwoods, and brushy scrub are all fine. Split cordwood works great.

Waste fuels often make great rocket fodder. You can use arbor and orchard wastes, dead branches, coppiced wood (straight "suckers" from pruned trees), and lumber project scraps (but avoid any lumber that has been treated with drying agents, salts, paints, or varnishes).

If you want to get competitive about efficiency, track your fuel usage by dry weight, not volume. Pound for pound, oak and pine are roughly the same BTU value even though they are quite different by volume.

Tinder/light fuels: Use these sparingly to start fires; overuse leads to excess ash. Paper, grease paper, dry plant wastes like leaves, straw, corn husks, reeds, etc., are good choices.

Small fuel scraps: Any clean, dry, natural cellulose–type fuel can be burned. Most won't self-feed like straight log fuels. Chips, wood pellets, dried dung, nutshells, block scraps, pine cones, and bark can be burned alongside other fuels in a mixed fire, sometimes with air adjustment.

Tricky fuels: Fuel/air imbalances can cause smoke problems indoors and out, and maintenance nightmares with creosote in the pipes. If you have an excess of poor fuels, try mixing small pieces in with a larger amount of good, clean fuel.

Avoid wet wood, punky rotten wood, too-long pieces, very pitchy wood, wide flat pieces shaped like chimney dampers, and forked or knobby wood sections. All these are poor fuels for rocket mass heaters, but can sometimes be burned in small quantities when mixed with better wood.

Recommended local fuel sources:
_____
_____
_____
_____
_____
_____
_____

Our heater's best-performing size and type(s) of fuel:
_____
_____
_____
_____
_____
_____
_____

Annex 6" Rocket Mass Heater, photo by Kacy Ritter

Bonny 8" Convection Bench. All photos of this heater are from "How to Build Rocket Mass Heaters" DVD, used by kind permission of Calen Kennett and www.villagevideo.org.

Finished Cabin 8" heater with brass filigree. (Copper and brass do not emit heat as well as dark, weathered steel. This black-and-brass version gives better heat output than the copper cladding.)

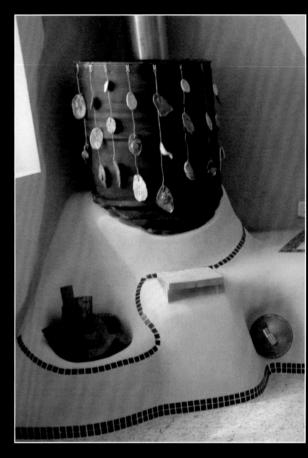

Detail: The barrel is decorated with ceramic bangles on stainless wire (photo by Adi Segal).

Top left: *Cabin 8" heater, in progress, with copper barrel cladding.*

Bottom: A mediterranean-climate design from Israel, as built by Adiel Shnior, Amit Pompan, and Nitzan Isrovitch in 2013. Photo by Adi Segal.

This greenhouse heater has a brick wall separating the firebox and barrel from the growing area. Brick stem walls, tiles, and wooden spreaders take the weight of small-scale aquaponic growing tanks. Photos courtesy of project owner.

Top left: *View from tipi entry, October 2015, courtesy of Priscilla Smith.
priscillasmithphotography.com*

Top right and bottom: *A rocket mass heater in a tipi – in Montana – a year-round glamor camping feature. While heating an un-insulated tipi (which is itself designed like a chimney, for summer ventilation) through a snowy Montana winter is far from sensible, the experience impresses visitors with the effectiveness of thermal mass heat storage.
Tipi and panoramic compilation photos copyright 2015, Arthur Held and Thomas Ferguson.*

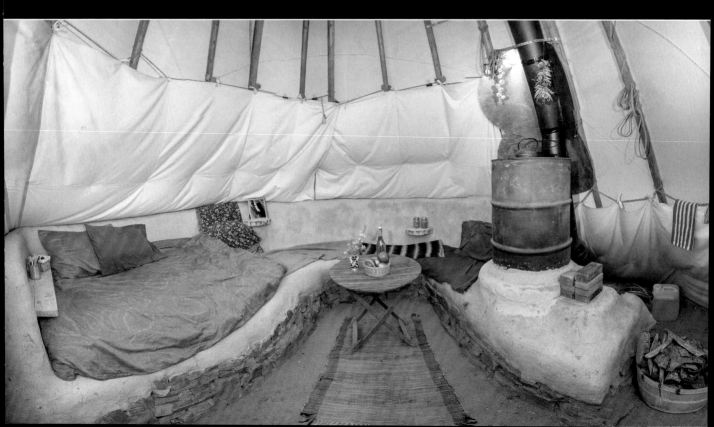

Wood floor is cut away leaving level, non-combustible supports. A bed of perlite-clay insulation is prepared for the firebox.

Construction views showing our example heater from Chapter 4 (also known as the Wofati 8" Heater). This heater was built during a teaching workshop, photos by author and students.

Brick firebox floor pad and bench pipes in place, including two of the cleanout access points.

Ceramic fiber blanket insulation, supported by wire mesh, surrounds the lower firebox. A modified elbow connects the cleanout T and the cut-barrel manifold, with a good 4" gap.

Firebox completed to the bridge course, and the manifold barrel is cut to fit the bricks and pipe

Heat riser and insulation, barrel, and cob. This is the sub-finish stage; now we let the core dry before adding tile or plaster.

After the first winter of use, the occupants reported they were able to keep this uninsulated log structure warmer than they had ever been able to do with their previous wood-burning stove, while using about one fifth the amount of fuel/wood. (Imagine how cozy it will be now that they've insulated the exposed walls!)

Top: *Finish plaster: an earth and fiber plaster, more comfortable and breathable than concrete.*

Bottom: *Wofati 8" Heater (as shown in step-by-step construction below). Photo by kind permission of Priscilla Smith, copyright 2015 Priscilla Smith, www.priscillasmithphotography.com*

Lighting the Fire

1. Inspect the heater and clean the firebox (leave ¼" of ash).

2. Use a candle or tinder to establish draft. Tuck lighted tinder partway into the burn tunnel. Draft is working when the flames pull away from you, toward the heat riser.

3. Drop kindling onto the burning tinder. Set the sticks vertically against the tunnel opening.

4. Once kindling is burning, add larger fuel behind the fire.

5. Load fresh logs behind the burning fuel. Re-load as needed, using the largest practical logs.

6. Adjust air flow if needed — at least ⅓ open throughout the main burn. A full fuel load and full-open air allows the fire to self-regulate.

7. As the fire burns down, reduce air flow.

8. When the fire is done, close the feed opening.

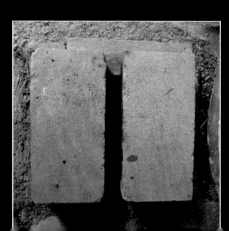

"Presto-Logs" (cakes of sawdust and wax) have not been tested; use caution.

BAD IDEA fuels: Some fuels can damage your heater and/or poison your household.

- *DO NOT USE* painted, varnished, or treated wood; plywood; or pressboard.
- *DO NOT USE* concentrated fossil fuels such as oil or coal.
- *DO NOT USE* volatile/explosive fuels, trash, metals, or plastics.

Not all firewood suppliers know or care about dry firewood. If you find a reliable supplier with good facilities for producing dry, seasoned firewood, don't lose their contact info!

Performance Log

Tracking the performance of your heater over time is the most accurate way to estimate fuel needs.

Use this book, or your own notebook, to record the performance of your heater. Note the fuel type, weather and outdoor temperatures, length of burn, heater and room temperatures (especially nearby combustibles/cushions), and whether any smoke is observed in the exhaust. Record any problems or discoveries.

Changes in performance from one heating season to the next may give early warning of maintenance issues.

Log the performance regularly until you feel comfortable with the heater (1–3 months). After that, you can log intermittently, a week or two at a time, to suit yourself. We sometimes log fuel weights and temperatures to check for performance changes, compare a new fuel source, or document our fuel needs in different weather conditions.

Maintenance, Inspection, and Repair

Over time, excess wood ash can build up into a hard mass, reducing the air volume and choking your heater. Before you start a new fire, scoop out the accumulated ash from your burn tunnel. Leave up to ¼" of loose ash, to insulate the fire and protect the bricks.

Tools for scooping out firebox ashes: hands, sardine can, bent-handle shovel.

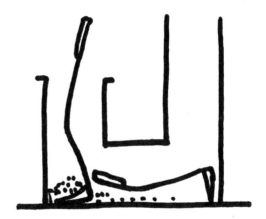

Annually/As Needed

Fly ash builds up like snow in the manifold and downstream, especially from burning paper. Cleaning time may be indicated by changes in performance: slow draft, weird slow burn, increasing backdraft/"smokeback," changes in air settings or burn times, or the sound of scrabbling vermin which must be removed promptly.

We recommend checking for excess ash after each 1 to 2 cords of wood (2 to 3 tons), or sooner if burning a lot of scrap paper, clay-treated orchard waste, or other high-mineral-content fuels.

The manifold/barrel base is a critical area to clean regularly. The nearby cleanout

Observation Log:

Date / Time	Wind/Weather	Fuel	Observations
2/1/08 6pm-9pm	NE / raining	21 lbs oak/cherry	:-) whole house is warm

Maintenance, inspection, and repairs _____

Maintenance: _____

Ash Cleaning: _____

Daily or weekly: _____

may be a capped pipe, door, or removable tile built into the barrel or brickwork. Some heaters are built to allow removing the barrel itself, or the lid.

- Protect nearby furniture if needed, then open the cleanout access.
- Remove ash and soot using a Shop-Vac with good filters, or a snake and brush. Inspect as you go for any damaged, sticky, or loose areas.
- Work your way down the heater, making sure that all channels are clear of obstructions. White ash or black soot are normal, and a small amount of moisture condensation may occasionally be present.

Ash Cleanout Locations

As a pass-down manual for your family and any future maintenance helpers, record the best access points for a person to inspect and clean all parts of the heater. Sketch access points, or include a photo showing the following parts: firebox, manifold, barrel, bypass, turns, chimney.

Ash Cleaning Log:

Firebox (daily/weekly):

_____ _____ _____ _____ _____ _____ _____ _____
_____ _____ _____ _____ _____ _____ _____ _____
_____ _____ _____ _____ _____ _____ _____ _____
_____ _____ _____ _____ _____ _____ _____ _____
_____ _____ _____ _____ _____ _____ _____ _____
_____ _____ _____ _____ _____ _____ _____ _____
_____ _____ _____ _____ _____ _____ _____ _____
_____ _____ _____ _____ _____ _____ _____ _____

Manifold/ducts (semi-annually):

_____ _____ _____ _____ _____ _____ _____ _____
_____ _____ _____ _____ _____ _____ _____ _____

Ash cleanout locations (sketch access points, or include photo):

Firebox _____
Manifold _____
(Barrel) _____
(Bypass) _____
Turns _____
Chimney _____

Pooling liquids, sticky tar, or enamel-like creosote indicate a problem requiring further work.

- Replace all caps and reseal any mortar or gasket joints that were disturbed, making sure no hardened residues remain that could prevent an airtight seal.

For removable bells and lids, use suitable, non-flammable seal materials such as:

- Braided or woven woodstove door gasket (seated metal lips)
- Mortar (clay-sand), cob, earthen or lime plasters (masonry areas)
- Foil tape*, furnace cement*, or chimney cement* (metal-on-metal)

*Check temperature ratings for suitability. Materials rated for 1000°F or higher may be used on upper barrel, but 300°F foil tape is only suitable for lower areas.

Inspection and Repairs

Any object collects a certain amount of damage through use and abuse. The most common repairs for rocket heaters are surface dings and chips, cracking or loosening of the bricks around the fuel feed, and upholstery cleaning and repairs.

Masonry Surface Inspection

Since the outer masonry is the only visible part, it is the first defense against leaks or cracking. Inspect before each heating season for cracking, damp, or other signs of damage, and repair as needed.

The most common surface cracking will occur within the first few months after an installation, and may indicate improper materials or methods (excess clay), settling in the foundation, or hurried work. We generally allow the core to dry completely before beginning the outer casing, so that any cracks can be repaired or stabilized and the final layers protected from associated stresses.

Cracking that occurs years into the useful life of the heater may indicate a more serious problem, such as a heavy impact or eroded foundations.

All surface cracks of unknown extent should be treated as critical to the air seal of the heater until proven otherwise. At a minimum, wet the earthen materials and press down to fill the crack. Repair methods for traditional masonry are described below, and more detail and training is available through the resources listed at the end of this chapter.

Upholstery

Never use combustible materials around the combustion unit and manifold, or anywhere that surface temperatures routinely exceed 120°F.

Needs repair.

Areas of the heater that remain below about 120°F (50°C) can have wood trim, cushions, blankets, or other comfortable padding added. Monitor closely during the first few burn cycles, though, and again during any unusually long burn, to verify that the temperatures *under* the insulating cushions stay within acceptable ranges (under 150°F [65°C]). Place a hand or thermometer below the cushions; it should be comfortably warm, not painfully hot. (The insulating cushions or fabric trap more heat at the masonry surface, so the surface temperature increases when cushions are added.)

All fabrics and padding should be heat tolerant. Fabrics that can be ironed or tumble-dried on "Medium" or "High" settings are heat tolerant (up to 350° to 450°F for short periods). For the safest results, we prefer natural fabrics like denim, cotton canvas, wool, or linen, and natural batting such as cotton or wool. If these materials do overheat, they scorch rather than melt. But many owners have used durable synthetic upholstery fabrics. Avoid delicates like rayon or acrylic, and avoid spray treatments like Scotchgard.

The only reported problems with fabric or wood trim in contact with the heat-exchange benches have been localized "hot spots" where a heat-exchange pipe was less than 3" below the masonry surface and the heater was fired for an extended period over five hours. Some scorching of natural fibers, and melting of synthetic fabrics, has been observed in these situations. If your heater has any such trouble spots, you may be able to fix them with additional masonry thickness, or a pad of noncombustible insulation.

Follow fabric directions for upholstery cleaning. Most owners clean their rocket mass heater bench much the same as they do their sofa: a quick vacuum or sweeping under the cushions and an occasional spot-cleaning, removing the cushion covers for a more serious washing if needed.

Bridge Brick

The most frequently replaced part of a rocket mass heater is the first brick in the burn tunnel. This brick is exposed to cold incoming air, very hot flames, and rough fuel handling. It may develop a vertical crack over time.

It is generally not necessary to replace the cracked brick unless any part of it becomes actually loose, threatening to fall into the burn tunnel during a fire. When it is time to replace this brick, soak down the masonry with water to reduce dust, and remove the bricks immediately above the broken brick. Carve away any loose material until you can remove the broken pieces and slide a whole, new brick into the same place. If there is a lot of wiggle room, reset the brick with clay mortar. Reset the bricks above in the same way.

Fixing It Better: Replacing a single, standard firebrick is a pretty simple job. Improved methods may offer slightly better longevity, if you wish.

To help protect from thermal expansion, you can bed the sides of this brick in flexible refractory wool or woodstove gasket. Paint-on refractory coatings may also help to reflect heat away from the bottom surface. To protect from careless fuel loading, it may be helpful to cut wood smaller or provide a slight lip sticking out above this brick to take the initial shock.

A metal air intake known as a "P-Channel" is popular with some builders; design details from Peter van den Berg are on the forums at donkey32.proboards.com.

Loose bricks can be reset using the same type of mortar. Clean off old mortar, chip back any protrusions, wet the surface, and reapply fresh mortar in a ridge or series of small lines. Set the brick back in place, and gently tap it until it is plumb, level, and square.

Masonry and Plaster Repairs

Among the critical factors in masonry repair are design for repair, structural integrity, and dust protection. Dust protection can be achieved with masks, curtains, and by wetting materials and tools while working.

Masonry repair is a skill that is best developed with practice. Basic tips are offered below, with more detail in Appendix 1. Local masonry contractors or a regional masonry heater builder may be your best

Bridge brick starting to crack, but not yet in need of replacement.

resource for renovations and repairs on conventional masonry.

Traditional Earthen Masonry

Many rocket mass heater builders use the original masonry heating materials: local stone, earth, clay bricks, and clay-bound masonry. Compatible finishes include earthen plaster, lime plaster, tile with breathable natural grouts, and other breathable non-synthetic plasters.

DO NOT USE vapor-proof finishes like latex, stucco, or oil paints over earthen masonry.

When working with local materials, especially site-sourced mineral soils, record the source details, amendments, and any other useful information (such as the phone number for delivered materials).

Repairing or Re-shaping Earthen Masonry

For deep repairs, to anchor new materials, or to replace a sculpted detail such as an arm rest, wait until the heater is cool (ideally during warm weather). Prepare new material with plenty of aggregate to prevent shrinkage cracking — test bricks are a very good idea, especially if you have not done this before or are using a different batch of materials.

Thermal Cob/Clay-sand Mortar

- Two to five parts* sharp sand
- One part clay-rich subsoil dirt*
- Water as needed to make a stiff dough-like mixture

*Depending on the local geology, we may use straight local mineral soils, rock-crusher fines, or sandy blends from 1:1 up to 10:1 sand:clay. Good garden soil is bad building material: avoid silty loams, topsoil, and organic matter (humus) as much as possible.

Mix thoroughly with a kneading motion, usually accomplished by rolling and treading on a tarp. For structural areas, add straw, hair, or other fiber as needed (as much as can be completely mixed in without clumping).

Testing Cob (Wet-formed Earthen Masonry)

A handful of good cob has a solid, crunchy center; excess clay or water does not goosh out when it is squeezed. You'll know it is the right consistency when it does not change any further as you mix it by hand.

There are a few tests you can make to make sure your cob is ready:

- Ball drop: holds together when dropped from 4′ height.
- Smear: sticks to lath or fingers, even upside down.
- Stand on it: does it make you taller? When the material is stiff enough, it may build up on your boots to 1″ or thicker, and you can stand on the pile like a mountain goat. Too-wet material gives way like a mud puddle; it can be laid out to dry for a while, and used later.

Once the cob passes the above tests, you'll test the cob's texture by making sample bricks of several proportions to find the best mix (shrinkage or cracking indicates excess clay, crumbling indicates silt or sand, and both together indicate not enough mixing).

Small amounts of leftover material may be dried and stored for later repairs. To re-use saved material, moisten the dried material in a bucket or basin (without too much excess water) and work it until pliable. Too much water can wash away clay, weakening the material. Work any excess water back into the mix with additional dry material.

Work the material until it reaches a dough-like consistency. For plaster and mortar, work in enough water to make a thick batter or frosting-like texture. For structural repairs, stiffer material (like shortbread or cookie dough) will fill bigger gaps, and shrink less.

Repairing Cracks

Before repairing a cracked or broken section, stop to consider the cause of the cracking.

- Does the crack open wider each time you fire the heater, then shrink as it cools, indicating a need for a thermal expansion joint? (See below).
- Does the area have some sculptural detail that is inappropriate to handle ordinary use and abuse, perhaps calling for a modified design?
- Was the original material too clay-rich, and thus shrank as it dried? Make sure the new mix has a lot more sand or fiber to reduce shrinkage.

Once you have determined the cause, gather your materials to repair the situation. Wet down the cracked area, and use a stiff brush to remove any loose material. A simple ding may just require minimal dusting off and a tiny patch.

Very deep repairs or alterations may require overnight soaking: put damp towels on the area, draped into a bucket of water and/or covered in plastic. Dry cob is almost as hard as concrete, and it gives off large amounts of dust during demolition, which are not good to breathe. Wet cob is much easier and safer to remove. If you can't avoid making dust, use good ventilation and/or respiratory protection.

Once the material softens, start carefully digging into the masonry with a garden or masonry tool until you get to the trouble spot. Use small strokes to avoid damaging pipes or other buried parts. Repeat the soaking and digging as needed.

When you have removed the desired amount of material, leave the surface rough to accept new material. You may wish to carve out jigsaw-like keys to let new material interlock with old, or tapered surface joins (like scarf joints in woodworking) to increase the surface connection. Straight, perpendicular joins are more likely to crack again.

Paint the areas to be joined with clay slip (a mixture of clay and water, about the consistency of paint or pancake batter). Work your new material into the repair zone. Use firm, pushing and smearing motions (don't pat; it makes a weak join that will fall off again later). If repairing a large section, build ledges like masonry courses, and tie into the surrounding material. Stop the rough cob at sub-finish level, with room for plaster. Smaller cracks can be repaired directly with leftover plaster from the original batch, for better color matching.

Feather the new material out along the old surface to blend in the repair. Finish the surface with a steel float, small section cut from a yogurt lid, or other smooth tools to match the rest of the surface.

Thermal expansion cracks expand when the stove is hot, then shrink as it cools.

The inside of the stove naturally gets hotter and expands more than the outside. If there is no space between the two layers, the outside will crack when the inside expands. You can recognize thermal expansion cracks because they get larger when the stove is hot, then shrink when it cools.

For small cracks, you may simply be able to patch the crack while the heater is hot. Fill the expanded crack while it's at its widest, forcing sandy plaster or cob into it, and keeping the heater hot until the filled area dries hard.

If cracking persists, you may need to add a flexible expansion joint (made with refractory insulation or a cardboard-filled air gap) to give the inner parts room to expand. Metal-to-masonry joins are particularly prone to unequal expansion. You can either remove the original casing to install the expansion joint and then rebuild it, or just build up around it for a larger final mass.

Lay a piece of flexible refractory insulation, or woven fiberglass gasket, around the corners or edges of metal and bricks that will get hot, to make room for these elements to expand and shrink. Build up the outer masonry casing around these materials, taking care to preserve the flex space between core and casing.

When building with a good fiber-rich cob, we just fire up the heater several times while the cob is still pliable, using thermal expansion to set the finished dimensions. Most masons would consider this cheating, but it works for us. If we are not able to wet-fire the heater (for example when using refractory cements that require a cool curing time), then we need to provide expansion joints around the manifold and firebox areas.

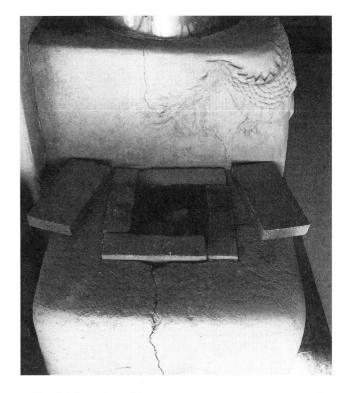

Repairing Cracks — photos courtesy of heater owners.

Redecorating or Refinishing

Sometimes you just have too many dings and scratches, and run out of matching plaster biscuits to repair them. Or you're inspired to try a fresh new look. Or both.

Refinishing Earthen Masonry

Permeable surface finishes like earth or lime plaster can be refinished with a similar plaster, or with compatible, breathable natural plaster or natural paint, such as clay-paint, lime plasters or whitewash, milk paints, or diluted water-washable paints such as egg tempera. (Do not use latex, acrylic, or oil paints: they are not breathable enough. If they don't peel off, they can trap moisture in the heater and cause deeper soft spots and blisters over time.)

Paints can be applied directly onto the dry surface, but may tend to wear badly in seating areas. With earthen masonry, just paint one coat at a time; let it dry; then add another coat later if you want more coverage. Reworking areas where the paint is still wet can cause the brush to pick up the material from below, creating discolorations or peeling. Brushes work better than rollers.

Plasters give a more durable coating, and also offer the option of adding some tile or sculptural detail for a bigger style change. Scratch the surface up before plastering, to a depth similar to the thickness of new material you want to add. (If someone has used a lot of oil and wax in the previous finish, you may need to "sand" the surface more deeply to remove the oily layer. You can make a giant sanding block for earthen masonry by wrapping expanded-metal mesh around a 2 x 4 block of wood, and adding a sturdy handle.)

Before applying plasters, prime the roughened surface with clay slip (using a stiff paintbrush allows you to remove loose material at the same time). Plaster is best

---

Material sources for this heater: _____

_____

Local earth/clay source: _____

_____

Local sand/gravel source: _____

_____

Other material sources (ceramic clay, mineral pigments, etc.): _____

_____

Samples/patch material: _____

_____

(Where do we keep them?) _____

_____

worked wet onto wet. Small tiles (no deeper than the plaster layer) can have their backs coated with clay slip and be worked into the wet plaster for mosaic or harder-wearing surface areas.

Work the plaster firmly into place with your hands or plastering tools. Create a smooth surface with steel floats or plastic tools, keeping the tools wet while working. Once the material is leather hard, you can burnish to a semi-gloss or even glossy finish if you like, using steel floats or a glass bottle or smooth stone.

You can do a clear or tinted "varnish" over these materials with soap paste (mix hot water with soap until it resembles the scum in the soap dish), milk paint or egg tempera, or a light coating of a drying oil such as linseed or poppy seed oil. Dilute the materials well, and work them in well,

Batch Recipe Notes

Record amounts and ingredients used for locally sourced materials:

Mortar(s)	Thermal Mass/Sandy Cob	Structural/Straw Cob

Insulation(s)	Finish Plaster(s)	Other

to avoid peeling or spotting. Bear in mind that clear coats will change the surface color, usually making it darker, and will also make it more work to replaster again later.

See also Appendix 1.

Troubleshooting

The most common or easiest solution is given first, followed by the increasingly "bad news" solutions.

1) *My heater won't light! (or) The fire burns straight up into the room, or goes out.*

Operation:

See operating instructions, including how to size and prepare dry firewood, and how to prime or check draft before starting the heater. If the mass is cooler than outside air, see cold start instructions, above.

Maintenance:

Clean the system thoroughly to check for blockages.

- Clean ash out of the firebox and manifold, leaving ¼" of ash on the firebox floor.
- Clear all channels of blockages (nesting vermin, paper ash); probe hidden areas.
- Remove the barrel or lid to check for blockages or ash buildup, including on top of the heat riser. Check the heat riser insulation and manifold for damage.

Construction:

Check for building/chimney problems such as:

- negative pressure: open a window (downstairs and upwind); turn off vent fans; close any upstairs windows
- blocked chimney? Clogged spark-arrestor screen?
- correct chimney height and temperature. An outdoor chimney which is cool to the touch, or shorter than nearby roof vents, may draft downward or stall.

Consult a reputable builder for help checking the system design and dimensions, and verify its suitability for your building situation and climate.

2) *My heater started fine, but now it's letting smoke back into the room!*

See above operation and maintenance tips, then consider the following:

- Water condensation in the exit chimney can cause chimney stalls (even very dry wood gives off more than half its weight in water when burned). Make sure the exit chimney is warm, as well as the firebox and mass.
- Wet fuel will steam and boil out the ends. Use only properly dry, seasoned fuel.
- Is fuel shorter than the firebox feed, so you can operate the air controls effectively?
- Keep the firebox full of wood. Set the air-control opening in proportion to the fire intensity. Most heaters operate best with the air between 100% and 20% open; leave at least a 1" slot open at all times while the fire is burning.
- Actively monitor the fire for the first 20 minutes, until everything is up to temperature. Fuel may wedge in the feed instead of dropping properly into the fire. Rattle the kindling to drop

any half-burned pieces, and load more wood behind, big ends down to avoid wedging.
- Watch odd shaped wood. Sawn lumber scraps can become a "damper" (wide, flat pieces placed to block the air) or "chimney" (isolated vertical channels between hot pieces of wood). Use natural rounds or splits to separate sawn pieces and keep the air flowing evenly downward between all the fuel.
- Piled-up ash, embers, or charcoal can block the burn tunnel. Bore an air hole through any such blockage with a poker or stick. To avoid this problem, clean ash to ¼" before lighting the fire, and use moderate-sized fuel once the fire is established (wrist-sized or larger).

If none of the above actions fixes the problem, you will need to consider somewhat larger issues.

Maintenance: Is it cleaning time? After the fire is out cold, CLEAN EVERYTHING (inside and out).

Construction: Certain fundamental errors can cause the heater not to draft properly.

Stagnation: If smoke chokes the fire about 30–45 minutes after lighting, a problem in the combustion unit may be causing the exhaust to stagnate as the heater reaches full temperature. Air leaks, inadequate insulation on the heat riser, improper barrel coverings (too much insulation near the outside of the barrel), or over-ambitious heat extraction can all cause this problem. Remove the barrel, insulate the heat riser and correct any other problems you may see (blockages, damage, air leaks), then reseal the manifold. A fresh interior coating of mortar or earthen plaster can quickly reseal the manifold, easier than hunting down individual leaks.

Negative building pressure: Tall buildings with leaky upper floors, or unbalanced mechanical vents in any building, can draw air and smoke backwards through the heater. Open a window/door on the upwind side of the building. Shut any upstairs windows. Turn off exhaust fans, such as kitchen or bathroom vent fans. If adjusting the available air fixes the problem, the permanent solution may involve balancing the home's ventilation with a heat-exchanging air supply, and weatherizing *upstairs* to reduce hot air leaks.

Chimneys: Is your exit chimney warm, tall, and well insulated from where it exits the building to well above the roof ridge? Chimneys near or below the roof eaves are subject to incredible wind eddies, gusts, and pressure disadvantages. If wind affects your heater performance, consider a better chimney, or a bypass to heat up your chimney when needed. We strongly recommend locating the chimney inside the building for the majority of its length, to reduce problems with heat loss to outside air. Exits near the roof ridge are often the most successful performers, and easiest to protect from leaks.

Proportions: Is your firebox correctly proportioned and insulated? Does the manifold provide a generous flow area from barrel to duct, with some extra room for fly ash? Have you avoided weird turns, bottlenecks, corrugated sections, or mismatched ducting anywhere in the heater? Improper proportions often cause performance problems.

3) *My heat-exchange mass leaks smoke, or smells funny.*

New heaters: Newly built heaters often steam gently until the cob dries out, which takes time. The smells involved are distinctive. Earthy, damp, or mild "barnyard" smells are normal, depending on the ingredients used. Smoke or mildew odors suggest a problem.

Cold smoke: A building with negative pressure problems can draw smoky smells backwards down any chimney, especially when the chimney or heater is cold. (See "Cold Starts" above, and "Smoke and Depleted Oxygen," below.) If smoky smells come from the heater's feed opening when the heater is not in use, see "Negative building pressure" and "Chimneys," above.

Leaks: If smoke is seen or detected coming from the body of the heater, ventilate the room well with outside air, and let the fire continue burning slowly while you try to find the source of the leak (check the barrel, manifold, and cleanout areas). After the fire is out, inspect for signs of leaks: black soot streaks outside the ducting, or bare traces of air leaks in the soot on the inside of the channels.

Most air leaks occur near exposed pipe such as cleanout caps. Use woven woodstove gasket to line any leaky connections between metal parts. Patch masonry leaks by removing loose material, wetting, and reapplying a fresh coat of mortar or plaster.

Sometimes it's easier to replaster or repoint the casing masonry, to create a secondary seal, if there are small leaks that are hard to find. See "Refinishing Earthen Masonry" above.

Mildew and mold: Damp conditions with inadequate ventilation encourage mold growth. In dry conditions, clay-based masonry remains dry and hard, and helps regulate indoor moisture by "breathing" slightly. However, damp problems can occur in any building.

First, remedy the damp problem. Every building needs adequate roofing, damp-resistant footings, and good drainage away from the building site. Look for blocked gutters, improper foundation details, building materials in contact with ground damp, leaking pipes, improper chimney or window detailing, condensation of warm moist air in unventilated spaces or against cold membrane barriers, or any other source of damp. Use locally-proven methods to eliminate the damp problem.

After creating dry, well-ventilated conditions, remove any remaining moldy or damaged material. Treat the remaining, structurally sound material with 3% hydrogen peroxide and borax/water solution to destroy and inhibit mold spores.

Repair the damaged area using compatible material.

4) *How do I take something apart to clean it, repair it, or redecorate the heater?*

See "Maintenance, Inspection, and Repair," above.

5) *I'm tired of this weird mud bench/big steel drum in my living room. How do I hide it, make it look better, or get rid of it?*

The bench and heater are a unit. Substantial alterations to either part can make it not work. Surface repairs and refinishing are described above under "Maintenance, Inspection, and Repair."

The barrel can be painted, decorated, or hidden behind a suitable heat shield such as fireplace screen, ceramic tile, or decorative metal panels. Or it can be replaced with a custom metal bell. See the Barrel Decorating Design Challenge in Appendix 6.

We are sometimes asked about eliminating metals entirely in favor of ceramics or other materials. To do this successfully requires expensive components and prototyping; consider a different, time-tested masonry heater design such as a contraflow heater.

To remove or replace the entire heater, follow the soaking and digging demolition methods outlined above under "Repairing Cracks," using a pickaxe or jackhammer in place of hand tools.

Safety note: Because rocket mass heaters exhaust at a lower temperatures, the through-roof or through-wall fittings used by an RMH installer may not be suitable for a woodstove or other appliance. Please have a qualified person inspect before re-using an existing chimney.

6) *We changed the heater, and now it doesn't work right. How do I make it work?*

   a) Change it back.
   b) Check whether a proposed change falls within the workable parameters for this type of heater as described in Chapters 4 and 6. We recommend testing any proposed changes, one at a time, in a full-scale outdoor mock-up before trying them indoors.

The building also influences the heater's behavior; see Chapters 3 and 4.

7) *At first my heater didn't warm up, but now it's working again. What's going on?*

Thermal mass lag time: Mass is great at stabilizing temperature, either hot or cold; the same property means it doesn't respond quickly to changes. We call this long, slow power-up and power-down the "thermal flywheel" effect. This time lag works in our favor to hold heat overnight, and to provide cooling during the peak heat of summer days, but it's not always convenient when returning from vacation.

When starting from scratch to heat a cold heater mass, it usually takes a day, with one or two firings, to reach full operating temperature. The hottest surface temperatures may occur 4 to 6 hours after the hottest point in the burn cycle, so don't overdo it on the first fire of the season.

During the heating season, masonry heaters work best when used routinely (several times per week). Avoid letting the heater cool completely until you are ready to switch over to summer cooling function.

Weird draft changes: Over time, ash can block parts of the system, and require cleaning. Vermin sometimes build nests in heaters if the exhaust is not properly screened. Weather affects draft too: most heaters draft best when the mass and chimney are warmer than outside air, so a cold day usually gives better draft. A heater that is optimized for winter draft balance may be difficult to start in summery weather.

Fuel changes: If you've switched to lighter or wetter wood, expect to get less heat from the barrel and heater.

8) *There's smoke coming out my chimney, or soot in the pipes. Is that bad?*

Smoke, or fog?

Rocket mass heaters should not smoke during the main part of the burn cycle, except possibly for a few minutes after loading cold fuel.

White fog or clear steam can be part of a normal, clean burn. White fog that is mostly water will dissipate (dissolve and disappear) as it leaves the chimney area — rarely more than 20 feet of plume, no lingering smell.

Smoke is usually colored (blue, grey, black, pale whitish, or even brown). The color remains in the air as the plume moves away from the chimney. It may spread out, but does not disappear. Smoke has a distinctive smell, and may make eyes sting or water.

Smoke often deposits inside the heater as black soot. A small amount of dry (not sticky) soot is normal, though less is better.

Smoke Diagnostics:
1. Dry wood is essential. Never burn wet, frozen, green, or unseasoned wood. Changes in fuel may require some practice to find the right fuel and air mixture to prevent smoke.
2. Creosote (sticky, tar-like deposits) in the pipes indicates an ongoing problem. Something may be blocked or damaged, reducing draft and air needed for clean combustion.
3. At the beginning of the fire, use very dry kindling and fill the firebox. At the end of the fire, reduce the air flow to keep the coals warm enough to burn clean.
4. Over-draft can cause incomplete combustion in heaters that are too short horizontally, or have a too-tall heat riser or exit chimney in a cold climate.

Self-Training: Good operators go outside to check for smoke from the chimney periodically, to train themselves on the most smoke-free fire-tending techniques for their heater, and to adjust those techniques for new fuels, weather, or firing patterns. With careful practice, it is possible to burn a smoke-free fire in almost any firebox; the J-style rocket just makes it significantly easier.

9) *My heater doesn't behave in _____ weather conditions.*

A heater that doesn't behave in ordinary weather conditions, like your prevailing winter winds, needs to be remedied.

Inadequate chimney? See "Chimneys" and "Negative building pressure," above. Chimneys that exit below a nearby roof eave or building will almost invariably experience gusts from the wrong direction in some weather conditions.

Draft balance is also affected by the length of thermal mass and the resulting exit chimney temperature. If your chimney is too cool to produce strong draft, it will be more affected by adverse weather. See "Cold Starts" above — consider a bypass or by-the-bell exit chimney, or a less ambitious thermal mass.

A few climates have truly weird weather, like sudden downdrafts during/after thunderstorms, which rattle the walls and may push air backwards through almost any chimney system. Thankfully, such extreme weather is generally rare and brief. If possible, let your weather-eye dictate the best time to operate the heater, so you can coast through the storms on stored heat. Start-up and shut-down are the most delicate times for the stove's draft, so the next best option may be to keep the fire going as normal, and ventilate the room if there's an unavoidable puff of smoke.

10) *Our cob cracked a lot as our heater dried out. How do we fix it?*

Why are cracks forming? Widespread cracks may indicate too much clay in the mix, causing shrinkage as it dries. Cracks that swell and contract near the hottest points of the heater indicate a thermal expansion problem. Random cracking or detachment of parts of the project may indicate poor technique (loosely patting materials into place, instead of working them into solid contact with the surfaces below). See "Masonry and Plaster Repairs," above.

Smoke and Depleted Oxygen:

What's the problem with a little smoke in the house? I like that "wood-smoke" smell. But avoid breathing smoke or combustion exhaust. Even invisible "clean" exhaust contains some reactive molecules like carbon monoxide (CO). These toxins can build up in the blood over time, impairing the body's ability to absorb fresh oxygen.

Symptoms of CO exposure or depleted oxygen include fatigue, irritability, headache, nausea, lethargy, dizziness or faintness, passing out or being hard to wake up. If anyone experiences these symptoms, ventilate the area with fresh air, and get everyone outside. (You will effectively get stupider, and sleepier, with prolonged smoke inhalation.)

Never leave the fire unattended. Don't "store" smoke in any heater, or attempt to shut down a fire while unburned fuel remains. Allow the fire to burn out completely, with adequate air, before shutting the heater down at bedtime.

One of the benefits of thermal mass heat is that you can freshen the air as much as you want, without losing your stored heat. Clear the air as soon as possible after any smoke exposure; even if the outdoor air is well below freezing, the mass will soon warm the room again.

Unproven Design Disclaimer: This book cannot cover all possible variations of wood-burning heaters, or the consequences of improvised design "improvements." New designs need careful prototyping and calibration. Sharing the information we've gained from our experiments does not make us responsible for others' work.

Resources

The performance of a rocket mass heater reflects its builders' skill and intentions, the operator's attention to fire tending and maintenance, and the knowledge available when it was built.

For more details, read the rest of this book and consider looking through the original *Rocket Mass Heaters* by Ianto Evans and Leslie Jackson.

Ongoing research and discussion may be found online: www.rocketstoves.com, www.permies.com, www.ErnieAndErica.info

(Note that www.rocketstove.com (singular) is a reference for smaller, lightweight cooking and portable "rocket stoves," and at the time of this writing does not offer current information on rocket mass heaters for thermal-mass heat storage.)

Methods and instructions for working with earthen masonry can be found in Chapter 4 and Appendix 1 of this book. You should also check natural building references such as *The Cobber's Companion, The Hand-Sculpted House, The Cob Builder's Handbook,* or *The Natural Plasters Book*.

Online references such as permies.com, practitioner websites, and members of the Cob Cottage Company, www.cobcottage.com.

Further Help

If you need help assessing or remodeling your heater, completing all the blank sections in this chapter will give you a very good place to start, before contacting outside experts for advice. Most of the current experts can be found through the above sources.

We recommend builders make at least two copies of their own manual for each heater: one for the builder to keep for future reference, and one to leave by the stove for all operators' convenience.

Chapter 6

Rules and Codes

Introduction: Types of Rules

THERE ARE SEVERAL KINDS OF RULES that we use when building a high-performance heater.

- Rules of thumb, derived from practice
- Natural laws and descriptive rules: "as a general rule, things work like this …"
- Legal rules: building codes, regulations, and local laws
- "Unwritten rules" of social/cultural norms (not discussed here, almost by definition).

Rules are human constructs: symbolic models of a situation. Real, natural forces don't always follow the rules in our heads.

In designing a heater for a household, we work to balance dozens of effects: gravity, convection, heat-induced thermosiphoning and stagnation; radiant heat transfer; conductivity and insulation; laminar flow and pressure differences; state changes like evaporation, melting, and condensation; stresses like compression, shear, tension, friction; uneven heat expansion; the whole complex chemistry of combustion; and the presence and effects of moisture on all these materials and processes. Next there's the human factor: resources, time, budget, tolerances and preferences, skill and interest, range of expectations.

Modeling the whole system in the abstract is almost as complicated as predicting the weather. So we fall back on empirical testing: change one thing, see what happens, learn and try again. We look for useful rules in related industries — HVAC, masonry heating, fluid dynamics. We look for patterns to define as rules of thumb, but we don't always know what's relevant. One builder may develop a rule based on linear dimensions, another uses surface area, another uses volume. Over time, we discover which rules are most accurate and useful.

The design rules in this book are young — a few dozen years in the testing, at most. There is still plenty of room for improvement. There will always be refinements and exceptions to these rules. We've given some of the background math equations in Appendix 5, and in Appendix 3 we've

A Swedish kakkelofen built by Flemming Abrahamsson, photo courtesy Leslie Jackson.

A custom bell-style masonry heater with bench, built by Max Edleson, photo courtesy Firespeaking.com

profiled some successful projects that fall outside the scope of our general guidelines.

The legal rulings that affect these heaters, such as building codes and clean-air regulations, are also the result of fallible human effort to understand and mitigate serious hazards. Every person involved has their own biases, expertise or lack thereof, and interests. Most of the people who made these laws had never seen a rocket mass heater.

At the time of this writing, several jurisdictions have issued permits for rocket mass heaters, and several certified masonry heater builders are working with rocket-type designs ... but there is no universal consensus about how the various codes apply to the peculiarities of this design.

Historic Masonry Heating

Masonry heating has a long history. Floor-heaters and heated platforms such as the Chinese k'ang, Roman hypocaust, and Afghani tawakhenah are at least 2,000 years old. Royally-sponsored competitions in the modern era produced highly refined designs including German and Austrian kacheloven, Swedish and Finnish contraflow heaters, and a range of Russian fireplaces and masonry stoves. Masonry heater designs blend with masonry bake-ovens across Europe and Asia, warming both food and home. Each of these designs is worthy of its own book.

Before building any masonry heater, it is well worth studying both the time-tested designs available and the features and lifestyles that their designers wanted. David Lyle's *The Book of Masonry Stoves* is an excellent resource, as is the North American Masonry Heater Association (www.mha-net.org) or similar regional associations.

A classic Russian masonry heater with oven, cook-stove, and bed platform (the platform is not visible in this view; it's hidden behind the two visible faces of this heater, above the deep oven). This drawing is based on a contemporary project built by Pavel Kruglov; originally shared on Facebook.

Regional differences and cultural values may affect not only the design, but also the performance expectations for efficiency, operator skill, and safety. For example, is fuel efficiency a safety concern? Some German, Austrian, and sub-Arctic design standards *require* complete air controls at the top of the chimney for efficiency, which requires a vigilant operator to manually adjust them at the end of the fire. Improper use of a top closure damper can cause deadly CO poisoning, but in these regions inefficient heaters historically caused deadly winter fuel shortages: it's a choice of dangers.

In milder climates, coastal areas, and in recently settled areas accustomed to the modern abundance of natural and fossil fuels, sometimes local regulations may *prohibit* the use of a complete top damper or closer, with air controlled only from the fuel door/intake side. In North America, the most common type of pivoting dampers

A cosy contraflow heater with bench and oven, built by Eric Moshier, image courtesy SolidRockMasonry.com.

A soapstone tiled heater with oven, built around the Solid Rock Masonry Heater core kit by Eric Moshier, image courtesy SolidRockMasonry.com.

A corner contraflow heater with oven, built by Eric Moshier, image courtesy SolidRockMasonry.com.

are legally required to permit a minimum of 15% flow, to prevent CO poisoning of inexpert operators and their households. Designs built around these requirements emphasize safety and ease of operation, at some sacrifice of efficiency, control, and heat retention.

(Inefficient operators may also contribute more to local particulate pollution and global CO_2 pollution; this demonstrates another aspect of American building codes, which protect the occupants first, sometimes at the expense of their neighbors.)

A masonry heater is still an unusual device in most North American regions, and may not be "grandfathered in" to newer building codes, as is more common in Europe. Availability and cost of parts, combined with the shorter time frame that most North Americans expect to own a home, can make owners baulk at the price estimates for a full-scale masonry heater.

Rocket mass heaters were developed to address both cost and efficiency concerns, making the most of readily available, recycled, and local materials. They began their development with Ianto Evans in the 1970s, as an offshoot of efficient "rocket stoves" for primitive kitchens. About the same time, the EPA was created, and it began regulating woodstoves and some other solid-fueled appliances; also at that time, the masonry heater builders were drafting the first ASTM (American Society for Testing and Materials) standards for masonry heaters. As a result, many known, traditional masonry heater designs were used to develop the current codes, while others have been designed after the fact to meet them, but rocket mass heater designs are coming along from outside. This leaves them even more ground to

A Cabin Stove with sidewinder-style rocket firebox, cooktop, and bench, built by Max Edleson of FireSpeaking.com.

Builder Lasse Holmes working on the core of a small sidewinder-rocket with warming oven, cooktop, and thermal mass bench.

The sidewinder cookstove and bench heater, with a couple of happy campers, at Wheaton Labs. This rocket-without-a-barrel masonry heater is an ongoing design collaboration between Lasse Holmes, Max Edleson, Kiko Denzer, Peter van den Berg, and others.

make up in jurisdictions that are unfamiliar with the whole masonry heating concept.

For example, if you live in a jurisdiction where the local building officials have decided to only approve UL-listed appliances instead of learning to work with the IRC masonry heater codes and ASTM standards, and you want their approval, your project choices may be limited to one of two brands of masonry heater — from the companies big enough to have developed UL-listed core kits. In other areas, any certified heater mason may be able to sell you a kit, or plans, that meet code, and act as the legal supervisor for your DIY project. And there are still a few revolutionary places that follow the ancient traditions: home owners can build their own heaters to suit themselves.

Most rocket mass heaters to date have been relatively experimental, built by owner-builders in off-grid or non-permitted structures. Reports of their efficiency and function are mainly anecdotal from these sources. We have worked and corresponded with hundreds of builders to document the most consistent and reliable methods for building these heaters. A lot of experimentation is still possible, and, doubtless, improvements will be discovered.

The release of this *Builder's Guide* does not mean the journey is complete! But we have recently seen a tipping point where the accepted "good design" seems to have stabilized. New builders are more likely to replicate known errors than to contribute useful improvements. So it seems time to document our work to date, and work toward widespread acceptance of the most practical, affordable, and effective versions of this design.

Rules of Thumb: Builder's Guidelines

The rules of thumb presented here are the result of experience and successful guesswork. Breaking more than two rules at once makes it more difficult to diagnose any resulting problems.

Site-built heaters are complex, and they require skill and judgment from their designer/builder. Most owners are drawn to rocket mass heaters by their reported efficiency, but may not understand how that efficiency is achieved. If owners balk at something critical to the heater's function, such as proportions, size, mass, exhaust detailing, location, or dry fuel storage, they risk poor performance and may even compromise safety.

Design Constraints

Each type of masonry heater from around the world has its own carefully worked-out proportions and rules of thumb. Design changes outside these limits may have unintended consequences.

For any design modification not described here (or described only in the appendices), we recommend testing a full-scale prototype in a safe place before committing to any permanent installation. (Test well beyond "normal" use, ideally 3+ days in a row. A wet or leaky prototype may not represent dry/cured performance, and the point of prototyping is to discover any potential problems.)

Combustion Unit and Channel Proportions

The primary design constraints are with the cross-sectional area: Every channel in the system must provide the minimum cross sectional area for flow. The downdraft area inside the barrel and manifold must be *at least* the size of this area, but may be significantly larger.

Our 8″ inside diameter (ID) system has channels with a cross-sectional area (CSA) of about 50 square inches. The firebox channels match this CSA, as do all ducts and chimney parts (8″ ID). The bell interior and manifold connection areas must have at least this same CSA for realistic flow, though preferably larger in areas where turns or fly ash may cause flow restrictions.

Give special attention to these spots, using the 8″ ID example:

- *The firebox opening:* all brick channels (burn tunnel, feed, and heat riser) are 7 x 7.5 inches, or a similar rectangle such as 6.5″ x 8″ (the extra few square inches allow for ash and turbulent flow)

- *The gap* between barrel and heat riser (imagine this as a paper crown, 2″ tall by 25″ circumference)
- The slot-like opening from *barrel into manifold* (at least 5″×10″; 4″×13″; 3″×17″; or 2″×25″)
- The opening from *manifold into duct* (at least 4″ clearance in front of duct for good flow).

Watch the same trouble spots for a 6″ ID system, with a CSA of 28 square inches:

- The firebox opening is 5″×6″ or 5.5″×5.5″ (choice usually depends on brick dimensions)
- The heat riser gap is 1.5 to 1.75 inches tall
- The manifold-to-pipe transition needs to be at least 3″ wide (smooth, slot-like transitions could be 3″×10″, 2″×15″, or 4″×7″)

The firebox dimensions for 6″, 7″, 8″, and similar-sized systems can be generalized into a set of *Proportional Rules*:

- The (minimum) correct heat riser height is at least three times the height of the fuel feed and at least twice as long as the burn tunnel.
- Proportions are NOT 1:2:3; the burn tunnel would be too long. Correct proportions are 1:1.5:3+, or 2:3:6+.
- For our 8″ example heater, the sizes that fit this ratio are a 16″ feed height, a 24″ burn tunnel length, and a 48″ heat riser height.
- With a common 23″ x 34″ metal barrel, the entire combustion unit footprint is about 33″ wide by 48″ long. The height from level foundations to the top of the barrel is about 54″.

Heat-exchange Channel Length

For an 8″ ID system, start with 50 feet.

−5 for Turns: Subtract 5 feet for each right angle (90°) turn (adjustable elbow or capped T). For an 180° turn (down and back) we subtract 10 feet.

−10 for Unnecessary Roughness: Avoid corrugated material — it creates enormous drag. For each foot of corrugated material, subtract 10 feet.

+10 Good Chimney Bonus: Tall, warm, straight. Located near the bell; rises vertically inside the building before exiting near the roof top. Add 5 feet of horizontal length for every 15 feet of warm, vertical chimney, or a total of 10 feet for the average well-built chimney inside the average well-built home. (Low chimneys, those subject to cooling, any pointing horizontally into the wind, and/or exposed outdoor sections do not offer a bonus and may induce back-flow.)

−10 Hot Days/Cold Starts: A cold start is when the mass is colder than outside air (may occur in warm climates, long absences, or occasional-use buildings). Shorten these systems by about 10 feet, to about 20 feet total, or provide priming options as described in Chapter 5 under "Cold Starts."

For a 6″ ID system, start with 35 to 40 feet, and adjust the same way:

−5 per 90° Turn
−10 Unnecessary Roughness
+5/10 Good Chimney Bonus
−10 Hot Days/Cold Starts

Moderate Roughness

Brick, ceramic tile, adobe, and other materials have been used but are not as smooth as duct-lined channels. Consistent cross-sectional area and smooth flow (both turns and textures) may provide similar performance to our typical metal-pipe lined channels; for example, brick or adobe channels can be built just slightly oversize, with room to smoothly plaster the inside before capping the channels. Unlined brick channels might call for a shorter overall length, due to friction along the sides. A masonry heater rule of thumb is that horizontal brick channels can usually run about 20 feet overall length.

We do not currently have rules of thumb for proportioning "bell" (hot-air-trap) or "plenum" (wide horizontal cavity) designs; however, Peter van den Berg's surface-area calculations may be very useful (see Appendix 3, and ongoing discussion at donkey32.proboards.com). Masonry heater research and prototyping are encouraged.

Minimum Length: No absolute minimum. 15 to 20 feet is a relatively short system, with hotter exhaust. Under about 12 feet you may find the exhaust very hot, heat storage less, and draft overly strong.

Other Layout Priorities

- Cleanouts must be located to give access to all system parts with tools like flexible brushes or a vacuum hose:
  - at the bottom of each vertical drop chimney, barrel-manifold)
  - at each 180° bend, and as needed to reach/inspect multiple 90° or shallower turns
  - as near as possible to the manifold
- Airtight seals, including expansion joints where needed
- Materials, clearances, and thicknesses for safe temperatures and adequate thermal storage (usually 4" to 5" masonry thickness)
- Effective placement and exposure of firebox, bell, seating, and cleanouts within building.

Discussion of Rules of Thumb: Correct Cross-sectional Area for Gas Flow

Think of cross-sectional area (CSA) as the smallest piece of paper with which you could block the exhaust flow. The simplest workable method for rocket mass heaters is to use the same minimum cross-sectional area throughout all the channels.

The barrel and manifold area may be larger (but not smaller). A system CSA of 150% is a reasonable minimum for exhaust flow where it changes direction from the barrel to the ducting, to avoid restriction due to turbulence. Some heater builders think of the barrel as a true bell, in the masonry heater sense — a place for the exhaust to stratify and give up some of its heat. For bell-like performance, the target would be more like 400% of system CSA, which doesn't fit easily into a 23-inch-diameter barrel. We have not yet found an upper limit on the volume that can be tolerated in the barrel and manifold area, though an extremely large bell might in theory rob too much heat, and require a proportionally shorter bench.

Systems of 6," 7," or 8" interior diameter work as described throughout this book. Systems of other sizes have been constructed, but they don't always follow the same rules.

Smaller systems such as 4" do not tend to work as channel-type mass-heaters (too much surface area, slow/viscous gas flow). Larger systems (10″, 12″) produce more heat and stronger draft, and may need special materials or clearances to withstand the resulting heat. See Appendix 5 for more detail on flow and draft. For a comparison of CSAs and the dimensional equivalents that produce a particular CSA, see sidebar.

Nonstandard brick or insulation may also affect workable parameters. For example, we have seen better insulation drive up interior firebox temperatures, sometimes beyond the melting point of that insulation; and we have seen taller heat risers speed up draft and change the surface temperatures of the barrel exterior. If you want to change the design for any reason, test the new design outdoors or in a safe experimental space first, and expect to go through several rounds of prototyping before the new design behaves predictably enough for safe indoor installation.

Heat Riser and Fuel Feed Limits

Three times the feed height is the minimum height for the heat riser. (Its height should also be roughly double the burn-tunnel length.)

The heat riser can be taller — up to 4 times the height of the fuel feed. A 12″ fuel feed works fine with a 48″ heat riser, if you are willing to cut your fuel to 11″. A 60″ heat riser can work with a 15–20″ fuel feed.

Most people go for the minimum proportion 1:3 for more than one reason:

- to accommodate reasonable fuel lengths
- to keep the top of the barrel at a convenient height for cooking on it and cleaning it

Equivalent CSA for Gas Flows:

8″ diameter ID = 50 square inches (50.3)

7″ × 7.5″	6.5″ × 8″	5″ × 10″
4″ × 12.5″	3″ × 17″	2″ × 25″

7″ diameter ID = 40 sq in (39.5)

6.5″ × 6.5″	6″ × 7″	5″ × 8″
4″ × 10″	3″ × 14″	2″ × 20″

6″ diameter ID = 30 sq in (28.5)

5.5″ × 5.5″	5″ × 6″	4″ × 7.5″
3″ × 10″	2″ × 15″	

Note that the rougher or narrower the opening, the slower the flow, due to surface drag. Comparing the outer circumference with flow area gives a good sense of the drag. In each example, the first pair of dimensions will give best flow. For openings like a manifold or bell outlet at right angles to the prevailing flow, 150% of the system CSA may be needed to avoid flow restrictions.

Metric:

200 mm ID pipe, CSA = 300 cm² (314)

18 cm × 18 cm	15 cm × 20 cm
10 cm × 31 cm	5 cm × 63 cm

150 mm ID pipe, CSA = 175 cm²

13 cm × 13.5 cm	12 cm × 15 cm
10 cm × 18 cm	5 cm × 35 cm

- so the barrel can be lifted over the heat riser without hitting the ceiling.

The burn tunnel length is constrained by the size of barrel. The feed should project just far enough outside the barrel to allow for a full brick rim around the feed not overlapping the barrel, or about 4″ of masonry around the barrel to avoid cracking. Smaller barrels allow shorter burn tunnels, sometimes an advantage for 6″ systems.

Channel Length, Chimney Design, and Temperature

The length and height of the channels and chimney have to be considered together.

Heat-exchange channel length affects exhaust temperature — longer channels strip more heat. Exit chimney temperature and height affect draft — hotter, taller chimneys draw more powerfully. Draft affects the workable channel length — a powerful draft can pull exhaust through longer channels. (Of course, longer channels result in cooler exhaust temperature, reducing draft.) Lower-temperature exhaust is more efficient (less heat loss); higher-temperature exhaust provides more reliable draft in all weather conditions.

So how do we design the exchange channels and exit chimney?

An existing masonry chimney can be used if its flue size and condition is suitable; see Appendix 3. For new installations, we generally use a manufactured chimney. Rocket mass heaters with exhaust temperatures below 125°F would qualify as an LT (low-temp) exhaust, but we typically call for a HT (high-temp) exhaust, such as a Class A chimney. This allows margin for errors or overuse, chimney priming, and the possibility that a future owner may re-use the chimney for a woodstove.

Adjusting layout and length of the heat-exchange bench are the simplest ways to adjust exhaust temperature. The designs presented in Chapter 3 represent our compromise between Ianto Evans's original ultra-efficient designs for small cob cottages, and the current conventions for masonry heaters that can be operated in any climate or season.

The maximum working lengths given above for each system size will produce an exhaust chimney temperature of 100 to 150°F (surface of chimney pipe). This is hot enough to rise under cold, dry winter heating conditions (-30 to 50°F/-35 to 15°C, which is typical of the US/Canadian border, northern Europe, and inland Eurasia), but may not be hot enough for reliable draft in warm or humid climates (California, Gulf states, Israel, Australia).

Lower-temperature exhausts may be more fuel efficient, but have less draft and are more cantankerous to operate in warm or windy weather. When this disadvantage combines with cold-start conditions such as a cold chimney or when the mass is below optimal operating temperatures, you can have a very balky stove that wants to run backwards (or not at all). Especially for milder climates or second homes, we favor a shorter bench for more reliable draft.

European masonry heaters typically exhaust at 200–300°F, to avoid the dew point of water in all conditions. They rarely run more than 20 feet of horizontal heat-exchange channels. A rocket mass heater built to this convention, in the short- to mid-range for our parameters, will enjoy more reliable draft in all conditions, but may lose up to 15% of its potential efficiency.

Exit temperatures as low as 60–70°F have been achieved in small, special buildings. These temperatures are often below dew point, making the exhaust too dense to rise. A conventional chimney will stall with such dense exhaust, but some owners have been successful by opening the bottom of the chimney (outdoor cleanout) as needed to drain too-dense exhaust. This low an exhaust temperature can make a stove unreliable.

Sometimes people suggest adding an exhaust fan to move cooler exhaust. We don't like to use exhaust fans to compensate for unreliable draft. Fans are a complex moving part, require electricity (don't work in power failures), and the corrosive and damp environment of a wood-fire exhaust gives most fans a very short life. A broken fan blade in the middle of the chimney looks a lot like a half-closed damper: it's an obstruction that could further choke the stove. Fans also don't guarantee the appropriate draft, and may even worsen performance: wood fires burn at a variable rate; the natural draft of the heat riser draws more air when the fire is hottest, less as it fades. More air at the end of the fire cools it too fast, causing more carbon monoxide pollution. A fan doesn't adjust in the same way as natural draft. And if you're trying to achieve "perfect" efficiency, the electricity and embodied energy of the fan must be counted against any gains from lowering the exhaust temperature.

Is there such a thing as too much draft? Too much air moving out of the building reduces heating efficiency. Too much air drawn through the fire can cool it to the point of incomplete burn.

Fortunately, excess draft can be controlled with a simple feed lid (two bricks over the opening, with a crack in between; or a purpose-built feed lid. Common feed opening settings are 100% open when starting and loading, then somewhere between 50% open down to a minimum of 20–25%

Heat exchange examples

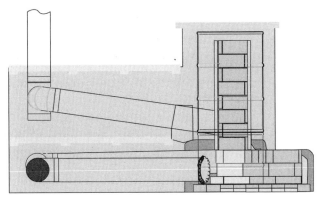

This fits right within the design parameters: duct length 20 feet, with 4 bends, bringing the effective drag to 40 feet. For a larger space, we could lengthen the straight part of the bench by another 10 feet (3 feet per section) and still be within cold-climate tolerances.
Result: Drafts reliably throughout heating season (when -30°F to about 60°F outdoors, with exhaust chimney surface during operation 115 to 150°F).

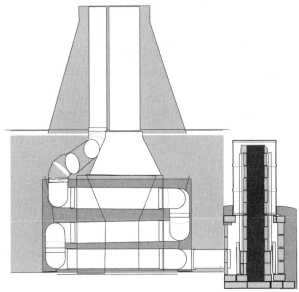

This example falls outside the design parameters: Duct length 23 feet, but with 14 elbows. This makes the effective drag over 90 feet. Result: This system requires chimney priming (using the built-in fireplace) before use. There is not enough draft to overcome the drag when the channels are cold. Though the owner is happy, we prefer not to repeat the experiment.

open for air flow during the fire. The lid can be closed down gradually as the fire dies, until it is just a crack over the last few coals, and then shut completely once the fire is out.

These heaters are much less prone to excess draft than a fireplace or stove due to their thermosiphon design acting like a U-trap when cold. However, they do continue to draft slightly on stored heat if not closed down, and are most efficient if closed after the fire.

A removable bell top allows for cleaning fly ash from both the manifold and the top of the heat riser. Households that burn more than two cords/year, or who burn recycled paper as tinder, may want a larger manifold ash-pit. Modern printing papers contain clay, and make a lot of ash.

Before you build, ask a local chimney sweep about their tools and space requirements for professional cleaning.

Insulation and Expansion Joints

The heat riser MUST be insulated for the rocket mass heater to function. In order for the thermosiphon firebox to work, the inside of the heat riser must remain very hot, and the outside (between heat riser and barrel) must remain cooler in order to flow downward. If temperatures in the barrel equalize (due to leaks, poor insulation of the heat riser, or inappropriate insulation outside the barrel), the draft stagnates and the system chokes and smokes.

Stabilized perlite or refractory wool insulation around the firebox doubles as an expansion joint, creating a "floating firebox" that can expand without cracking the casing. This high-temperature area is the most critical for expansion; the next most critical is the seal around the barrel.

Earthen masonry can be hot-fired while still flexible, to create the necessary expansion tolerance without cracking. (If cracks occur, earthen masonry is easy to patch, and can be hot-patched to spread the crack at its widest extent.)

For a project that can't be hot-fired during construction — for any reason — consider adding an airtight expansion joint around the barrel base.

There is more detail on expansion joints in Chapter 4, "Repairing Cracks," although if you are building with refractory mortars that

> **Conductivity and Insulation:**
>
> (Superconductors, molten sodium)
> - Most metals (copper, steel, lead, iron)
> - Water
> - Soapstone
> - Firebrick, fieldstone, concrete
> - Glass
> - Clay-sand mortars, most plasters
> - Earthen masonry (adobe, cob) with straw
> - Sand, loose dirt
> - Wood
> - Pumice, expanded clay
> - Straw-clay, sawdust-clay
> - Vermiculite
> - Clay-stabilized perlite
> - Loose perlite
> - Kiln brick
> - Rock wool/ceramic fiber insulation
> - Air (nonmoving)
>
> (Aerogel, vacuum (nothing))
>
> ↑ More Conduction / More Insulation ↓

require a long cool curing time, you may want to build in the expansion joints as prevention rather than try to patch cracks afterward.

Mass/Size Calculations

Predicting heat delivery is a rough science. Chimney height, fuel choices, operation and maintenance affect draft speed and burn rates. Building size, detailing, wind, weather, indoor and outdoor temperatures all affect heat loss. Even personalities matter; is the household willing to share the warm spots as needed, or must heat delivery be precisely "fair" to every distant room?

One base for comparison is fuel consumption. Complete combustion of wood at 0–20% moisture yields 6200 BTUs (70% efficient woodstove) to 8700 BTUs (perfect, ambient-temp exhaust) per pound of wood.[1] (Roughly 10 to 16 megajoules per kilogram.[2]) Incomplete combustion may release less than half the heat.

Our Cabin 8" heater may burn 30 to 40 lbs in 4 to 6 hours; so we can calculate an average for both values: 35 (lbs) × 7000 (BTUs) = roughly 250K BTU. A 6" heater might burn about half the wood. Some of this heat (up to about half) is released at the bell during firing; the rest is stored by the mass.

Some masons calculate heat output in terms of the surface area of the heater (inside for heat extraction, outside for rates of output into the room, and for heat loss calculations compare the surface area of exterior walls). Engineers may try to calculate the heat storage capacity of the masonry mass itself — although temperatures vary through the mass, the average temperature at the end of a firing cycle is somewhere between 500°F (core) and 120°F (surface). The mass delivers a quantity of heat into the room as it slowly cools to ambient temperature. Nearby structures such as concrete slab floors, dense partition walls, etc., may also heat up during the firing cycle and contribute to the overall heat storage potential of the heated room.

At the time of this writing, we have not worked much with these calculations, and encourage readers who prefer this method to use the heater examples from Chapter 3 to generate their own comparisons.

Heating Cycle Time Estimates

Masonry thicknesses should be AT LEAST enough for structural integrity: 4" thickness around all ducts, as for a lined masonry

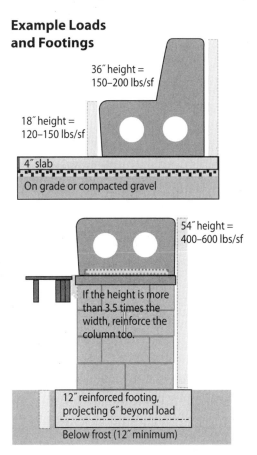

Example Loads and Footings

Common Errors

The following errors cause performance problems beyond the scope of this book. They would be considered "significantly not as described."

- *Avoid interior constrictions or bumpy added "features."* Turbulence or fly ash can turn a slight bottleneck into a major obstacle. Watch firebox, heat riser outlet, manifold, turns, and exit chimney caps for these common mistakes:
  - undersized slots or poorly flowing junctures (right-angle turns in the manifold may need to allow at least 150% of the CSA to avoid restrictions)
  - flat, shallow openings that get choked with fly ash
  - measurement errors ID/OD (inside dimension/outside dimension). This is particularly common in the brickwork (measure from inside corners, don't line up outside corners); the heat riser gap (measure from inside the barrel, it's nearly impossible to get an accurate measurement over the outer rim); and in insulated or double-walled sections for exit chimneys (look for ID on labels, or use your measuring tape on the INSIDE).
- *Improper dimensions in the firebox and overall proportions.*
- *Making the feed larger to take larger wood.* This doesn't work. Problems include smoke pouring into the room, creosote buildup in pipes, and possible flash-back fires in accumulated smoke (and a dirty burn which is less efficient, if you still care about efficiency at this point). If a larger firebox is needed, consider a tested design such as the batch box in Appendix 3.
- *Making burn tunnel too short*: Even if you compensate by making it wider, low burn tunnels get blocked more easily by ash or embers, and are harder to clean out.
- *Random substitutions:* Avoid unsuitable materials, or even "better" materials in the wrong places. We have listed suitable alternatives in Chapter 4 and Appendix 1.
- *Heat-exchange channels that are rough, corrugated, too long, convoluted, or mixed size.* These often cause extreme drag and poor-to-nonexistent draft. Stick with smooth materials of consistent dimensions.
- *Chimneys that are too large, cold, or exposed (for an exit chimney).* These situations cause intermittent, terrible draft. (Slow, cool exhaust gases can stall if water condenses in the chimney; the chimney may still be warmer than outside air, but wet exhaust is denser than air. Impact varies with weather.)
- *A freaky hot firebox.* A super-insulated firebox (with no dense firebrick liner) can drive fire temperatures above 2300°F (perlite and many other refractories will fail). The lack of brick may also lead to flame irrigation, abrasion damage, and eventual sudden failure. Partial liners may lead to uneven heat retention and invisible draft failure during cool-down.
- *The right stuff in the wrong place.* Failure to insulate the heat riser, or mistakenly

adding insulation on the outside of the bell, can completely stop the draft within about 40 minutes after lighting the stove. Air leaks anywhere in the combustion unit cause similar problems (including improvised "air intakes" or "outside air").

- *Metal is warping or burning.* While metal is considered noncombustible in ordinary use (such as cooking), the aggressive heat of these clean-burning fires will quickly warp or burn out metal components in the firebox (usually in less than one year). Metals also expand more with heat, acting as a slow hammer on nearby materials. Notice that the metal pipes and barrels in this system are away from high heat areas and direct flame exposure, rounded to protect nearby masonry, and either double-sealed or easily inspected.

Thermal Mass Example:

Building: 800 sf cottage, 60 cu. ft. heater. (24x24 foot concrete slab floor, drywall, insulation, etc.) January av. 23°F. Overnight heat loss: roughly 95,000 BTU at 65°F indoors; 75,000 BTU at 55°F.

Overnight, the heater's surface temp drops from 95–120°F to 75°F . . . call it 20 to 40°F?

Heat capacity of 60 cu. ft., with a drop of 40°F:

Clay, brick .22 to .24 BTU/lb/°F × 120 lbs/cf × 60 cf × 40 °F = 63,000 BTU

Concrete/stone .18 BTU/lb/°F × 150 lbs/cf × 60 cf × 40 °F = 64,000 BTU

Adobe (dry earth) .3 BTU/lb/°F × 95 lbs/cf × 60 cf × 40 °F = 68,000 BTU

These rough estimates may not be accurate or useful. A heat loss of 40°F seems big; we more often see surface temperatures drop 10 to 20°F at the heater, and 10°F (from 75 to 65°) in the rest of the house. Other heat sources (sun, cooking) and/or other storages (structure, furnishings) may contribute to our comfort.

(Sources: www.engineeringtoolbox.com, www.greenbuildingadvisor.com, www.builditsolar.com)

chimney. However, there are circumstances when you might want to adjust the masonry thickness around the pipes. It's your choice whether to include any footings, slab, or masonry hearth in the heater mass/thickness, or to try to isolate them with insulation.

Heat penetrates earthen masonry at about 1″ per hour. Thicker mass gives a slower and longer-lasting heat storage, but if it's too thick, it may never warm up to comfort temperatures. Theoretically, an 8″ thick mass takes a full day to warm up; but it can provide stable warmth for more than a day after a good fire.

Note that our squared-off mass with round pipes in it naturally has a range of thicknesses, which spreads the cycle out. Heat penetrates the front of the bench within a few hours of firing; the heat that migrates deep into the bench may not emerge until two days later.

The benches diagrammed in Chapter 3 generally give one to three days of heat storage, depending on outside temperatures. (The Daybed 6″ is rarely fired more than twice a week in the mild coastal climate where it is located; the Cabin 8″ is fired once per day in deep winter conditions and every two or three days in the shoulder season, to maintain comfort temperatures.)

Some builders have experimented with very low-mass heaters for occasional use, such as a covered outdoor entertainment area. One colleague uses fiber-reinforced masonry panels to support people's weights and lateral loads, with about 2″ thickness in the thinnest areas. These thin areas heat up more quickly for on-demand comfort, but may become painful hot spots if the heater is fired over an extended period.

Building Codes and Rocket Mass Heaters

After much research, we've cut a lot of detail from this section. The bottom line is, every jurisdiction can write its own rules. We are not lawyers or even licensed building contractors; our advice may simply confuse the issue when it comes time to work with your local authorities.

However, having done the research, we figured we could at least provide useful references, and comment on a few known issues. Codes designed for other types of solid-fueled heaters might mislead builders or officials into recommending inappropriate design changes, and we don't want that.

Applicable Codes

In North America, the most relevant code documents are the International Residential Code (IRC), section R1002 for masonry heaters; and the corresponding ASTM standard E-1602 for masonry heaters. (International Building Code also has a section B1002 for masonry heaters; however, balancing the chimney draft and heat output for commercial buildings over two stories tall is beyond the scope of this book.)

ASTM 2778 covers tile stoves specifically, but is not generally referenced in the building codes as is ASTM E-1602.

Masonry heaters traditionally are site-built, not manufactured and shipped. They historically have not been regulated by the EPA, like woodstoves are. They are exempt by weight. (By the ASTM definition, a heater must be at least 800 kg/1760 lbs, though some others give a definition of 1500 lbs.) Woodstoves are under 900 kg or 1980 lbs by EPA definition, so there is some potential overlap. The rocket mass heaters as described here would generally weigh upward of three tons (6000 lbs/3000 kg).

There has been some work on an EPA-approved emissions test for masonry heaters; however it currently applies only to batch-burn masonry heaters, and is not suitable for testing the J-style firebox. A jurisdiction that requires EPA certification above and beyond the requirements of R1002 may inadvertently exclude a wide range of clean-burning traditional masonry heaters, while allowing marginal performers such as small woodstoves that are more easily transported to testing facilities and can be tested under existing protocols.

The North American Masonry Heater Association membership lists can be found at www.mha-net.org; the Alliance of Masonry Heater and Oven Professionals, at www.masonryheaters.org. A certified masonry heater builder has some advantages when seeking official approval.

In Europe, the most broadly-used references seems to be the British-Adopted European Standards, BSEN:

- BS EN 13229 covers inset masonry heaters and open fireplaces burning solid fuels
- BS EN 15250 covers slow heat release appliances fired by solid fuel (includes some specs for manufactured and partially pre-fabricated appliances e.g. masonry heater core kits)
- BS EN 15544 covers kacheloven/grund oven (tiled/mortared stoves)
- BS EN 8303–1 covers heaters burning solid mineral fuels (coal)
- BS 8511 covers solid-fuel-burning heaters installed in small craft (boats)[3]

In other English-speaking countries such as Australia, New Zealand, and South Africa, the general approach seems to be to fall back on rules designed for chimneys, fireplaces, structural soundness, and fire prevention. Masonry heater builders are likely to be imported, or to travel internationally for their training.

The Germanic and Slavic countries have a tremendous history with masonry heating. We authors regret being unable to access most of this knowledge except in English translations such as: TRVB 105/86: Fireplaces for Solid Fuels (Austrian National Fire Service Association, OBFV, and Austrian Fire Prevention Services) found at www.bundesfeuerwehrverband.at

Those with suitable language skills may wish to look up the Austrian tile stove association, www.kachelovenverband.com, or similar masonry heater associations across Europe and Asia. (For Russian fireplaces try searching Русская печь; for Finnish versions, pystyuuni or kaakeliuuni, for Swedish, kakelugn or contraflow stoves; Chinese: k'ang 炕; Korean ondol or gudeul, 구들.)[4]

Design Changes, Authority, and Expertise

Local building officials are often selected based on knowledge and/or experience of the building industry. However, even a brilliant person may work in the building industry for several lifetimes without seeing or knowing every possible way to build things.

Local authorities may suggest design changes to accommodate existing rules about solid-fueled heaters. Making unproven design changes in order to satisfy an arbitrary rule is asking for trouble, because the writers of those rules generally had no experience with this specific type of heater. Unwarranted design changes may cause performance and safety problems beyond the scope of this book.

Anyone who has not build rocket mass heaters, yet who suggests design changes, is operating outside the scope of their expertise. (In fact, we'd suggest building five or ten rocket mass heaters before claiming expertise, and extensive prototyping of any proposed design changes before implementing them in an occupied home.)

We respectfully suggest that if these heater designs *as tested and documented* do not meet local rules, then the appropriate solution is either seeking a rules variance or amendment or considering a different, proven masonry heater design that does satisfy local rules.

Jurisdictions do have the authority to write their own rules and exemptions. Sometimes special rules get written that favor special interests, unexamined beliefs, or a temporary fad (such as the belief that natural gas is "cleaner" than renewable fuels … don't let us get side-tracked on that one!). If the jurisdiction realizes that it has unduly constrained itself or is working against local best interests with arbitrary self-imposed rules, the jurisdiction may wish to remove such amendments, or revert back to adopting the current IRC codes without amendment.

Some jurisdictions have adopted a specific exemption or appeals process to allow clean-burning rocket mass heaters through their own special approval process (see Appendix 2). While in general we'd prefer to see widespread acceptance under the existing IRC and ASTM standards, a local exemption or variance can be a workable stop-gap solution.

There may also be standing local exemptions that can be used for a particular rocket mass heater project, bypassing the more involved and costly appeals process.

Here are some questions to ask during an initial meeting with local building officials:

- What are the permitting requirements for a woodstove, masonry heater, or other solid-fueled heaters? What are the fees/other costs?
- Do we have any special local rules, like seismic zoning, air quality, or local chimney height standards that might affect my plans?
- Are there incentives for energy-efficient appliances that might apply to a site-built masonry heater?
- Do I need an engineer or pre-approval for my foundations? Or can I just look up the load in a table and use common sense?
- Are there any exemptions that might apply? For example, if I wanted to lay up my heater in a shed or greenhouse first to practice, could I do that without a permit?

Many areas allow exemptions:

- for buildings under a certain size, or not on permanent foundations
- by weight (heaters over 900 kg are not a woodstove under EPA regulations)
- if an appliance is the occupants' only way to heat or cook
- for approved experiments or prototypes; to be reviewed for permanent approval
- for antiques over a certain age, or of unusual design or value
- for small outbuildings under a certain size (varies by jurisdiction), off-grid cabins, or owner-occupied homes.

Acceptable Design Constraints

For reference, following are some code-required practices that have proven compatible with our rocket mass heater designs.

International Residential Codes

R1002.1 Definition: "A masonry heater is a heating appliance constructed of concrete or solid masonry, hereinafter referred to as masonry, which is designed to absorb and store heat from a solid-fuel fire built in the firebox by routing the exhaust gases through internal heat-exchange channels in which the flow path downstream of the firebox may include flow in a horizontal or downward direction before entering the chimney and which delivers heat by radiation from the masonry surface of the heater."[5]

A rocket mass heater would apparently fit this definition.

R1002.2 Installation: Offers a choice of complying with ASTM E-1602, or installing listed and labeled appliances (UL 1482) according to manufacturer's instructions. As of this writing, rocket mass heaters are not listed, and so would fall under the ASTM standard.

R1002.3 Footings and Foundations: The minimum firebox floor thickness of 4" (102 mm) is roughly compatible with our common building practice of a 2.5" layer of fire brick, backed with 1" to 2" of mineral insulation. (If we lay 2" of clay-stabilized perlite, we've met the requirement with some allowance for compaction.)

For noncombustible foundations, the masonry heater code refers to the masonry chimney codes, R1003.2.[5]

We agree with the concept of noncombustible supports, and there is no problem with building a rocket mass heater on top of a concrete slab if you happen to have one available. However, the 12" thick, reinforced concrete foundation requirements for full-height masonry chimneys (20+ feet tall) seem excessive for a masonry bench less than 2 feet tall. The ASTM standard gives an out, allowing for foundation thickness adequate to support the weight of the heater.

In practice, most masonry heater builders prefer to use the appropriate slab thickness to support the actual weight of their project, varying from 4" to 6" for small ovens and low benches, 8" slabs for modest-sized towers, and the 12" reinforced slab generally reserved for projects of more than one story in height (such as where a footing must be built up through a full basement to locate a heater on the main floor of the house).

The engineering calculations for the Portland approval process (Appendix 2) suggested that a 4" slab would be adequate to support earthen masonry benches up to 30" tall, or a shorter bench with a narrow raised back up to 48" tall. This agrees with our experience (we have observed at least two projects built on existing 4" to 6" slabs, over about 4 years in each case, with no evidence of cracking or shifting of the slabs despite at least one minor earthquake during the observation period). Areas with extreme frost heave may need deeper or heftier foundations.

A certified heater mason or licensed engineer's stamp on this section of your plans may be useful to mollify local building officials — if you can get it for less than the cost of the extra concrete.

R1002.4 Seismic Reinforcing: We have seldom built a rocket mass heater tall enough to require seismic reinforcing. The extreme improbability of the low bench ever falling on anyone can be seen as a reassuring safety feature.

R1002.5 Minimum Clearance to Combustibles: The general requirement is that combustibles will not be placed within 36" of the heater (as for non-certified woodstoves under many codes), except: Where the firebox walls are not less than 8″ of masonry thickness, and the heat-exchange channels not less than 5″ of masonry thickness, the acceptable minimum clearance to combustibles is reduced to a 4″ air gap.

We have built heaters with the 4″ air gap (see Bonny 8″, Chapter 3), and they work great. Eight inches of masonry around the firebox is very reasonable; five inches around the heat-exchange channels may be an inch or so more than necessary at the cool end of the bench, but is not unreasonable. (An experienced mason might recognize the 5″ standard as implying more than one brick wide; multiple layers reduce the risk of through-cracking at the joints.)

The 4″ air gap is a nice width for air flow, cleaning and retrieving fallen items, and controlling vermin. The 4″ air gap plus 5″ thickness agrees nicely with the 9″ minimum for HT manufactured chimneys (stovepipe with heat shielding, or double-walled pipe).

The ceiling clearance of 8″ from an insulated capping slab to the ceiling DOES NOT apply to our metal radiant bell. We recommend following guidelines for non-certified woodstoves and/or treating the metal top as a cooktop. It might be possible to reduce the clearance to 18″ with excellent heat shielding and/or insulation on the barrel top. However, for practical installation and removal of a standard 32- to 34-inch-tall metal barrel, 36″ from the ceiling is a nice clearance.

ASTM Standard E-1602 (2003–2010 Version)[6]

ASTM E-1602–03 section 5.2 Clearances and Heat Shielding: These requirements are harder to interpret, as they assume a vertically oriented firebox opening. Combustible walls must be shielded if they are less than 48″ from this opening, and the area directly in front of the firebox must have a 48″ clearance with no reductions. However in our case, the perpendicular direction from this opening would be straight up.

In general, we do recommend allowing plenty of room to operate the heater, and providing a spark-safe storage area for fuel conveniently nearby to reduce the temptation to store kindling or firewood right on the open hearth. For convenient operation, 48″ is not unreasonable space to have on one side or another, but it seems unnecessary to have 48″ in all directions. We suggest considering the 48″ requirement as pointing straight up above the firebox for radiant heat protection, or fanning out along lines drawn from the bottom corner of the firebox to the opposite edge. There should be no combustibles located directly above the firebox, nor on top of the barrel.

In general, we recommend a simple performance-based standard: If any combustibles near the heater are too hot to touch, additional heat shielding is needed. (The actual upper limit is about 150–165°F (65–75°C) — not more than 90°F (50°C) above ambient temperature.) Always install heat shielding with air gaps: not only is it more effective, but it's also easier to double-check its effectiveness and make sure your combustible surface is now remaining safely cool.

Our experience in shielding the metal radiant bell seems to agree with older clearance recommendations for non-certified woodstoves: a 36″ to 48″ minimum clearance without heat shielding, which can be safely reduced to 18″ to 24″ with good heat shielding (e.g. a tile, ≥ 24-gauge metal, or brick panel, with 1″ air gap to wall, and generous air gaps below and above, so that air can flow freely between the panel and the wall), or as little as 12″ to 16″ with excellent heat shielding (double-layer construction with tile or brick, backed with a continuous metal sheet, with the same 1″ air gaps behind, and larger air gaps top and bottom).

The smaller numbers in front (starting with 36″) are for wood burning stoves with shielding, and these distances seem to agree with our experience of the metal barrel rocket mass heaters (the metal is not the firebox, but a secondary layer). 48″ is for wood burning stoves without shielding.

More Details

The minimum legal clearances above which I have described as 1″ are actually given as ⅞″ (which allows for some slight variation in materials). The minimum clearance for air flow at the top is 3″, and at the bottom is between 1″ minimum and 3″ maximum. Ceiling shielding needs a minimum of 3″ edge clearance. These clearances must be established with noncombustible spacers (such as ceramic insulators or 1″-long pieces of metal plumbing pipe, which can be threaded onto the bolts or lag screws holding the heat shield on the wall). The fasteners must not be closer than 8″ from the vertical center line of the stove, and if they extend into combustible materials, must be at the lateral extremities (far sides) of the shield. The shield is to extend 18″ beyond the appliance on the sides, 20″ above the appliance on the wall.[7]

Masonry or noncombustible materials applied directly to a combustible wall may trap heat in hidden spaces, and are not considered to offer any significant protection from radiant heat nor to reduce safe clearances.

If in doubt, check the temperature of any combustible materials near the heat source (you can get quite reasonable point-and-shoot infrared thermometers at electronics or woodstove shops). All combustible materials should remain below about 150°F (65°C) throughout the heating cycle. If the wood and paper of your home is not staying at safe, touchable temperatures, improve the heat shielding, move the combustibles, and/or replace the wall with a noncombustible wall.

ASTM E-1602–03, section 5.4 Hearth Requirements: The standard describes four types of firebox opening (floor level and raised, various sizes) and prescribes 16″ or 20″ minimum hearth from each firebox opening.

Our J-style firebox opening is a horizontal hole, so the 8″ masonry thickness discussed above naturally makes an 8″ hearth. Adding another 12″ of hearth on the floor would meet the more generous 20″ hearth requirement, as measured from the firebox opening.

In our experience, it's rare but possible for popping sparks to escape the firebox. A 20″ hearth, as above, seems reasonable to protect combustible floors. The overall width of the hearth area would be about 48,″ centered on the firebox opening.

However, numerous builders have omitted this hearth, with no more spark damage than is commonly seen adjacent to a legal-sized fireplace hearth. It's possible that a listed rocket mass heater in the future might only require the 8″ minimum hearth naturally provided by the firebox masonry.

ASTM E-1602–03 5.2.5 Wing Walls: These seem clever, and likely to work with rocket benches. The wing wall is a small masonry wall, no more than 4″ thick, extending from the heater to meet nearby combustible partition walls at a safe distance from the heat.

We generally would NOT use masonry wing walls around the combustion unit. The radiant metal bell needs to shed heat for proper draft function, and its greater thermal expansion is likely to crack any masonry in contact with the upper portion of the bell (the hottest part). It's also nice to have the option of removing the barrel for inspections and maintenance down the road, for example to clear fly ash that may settle on top of the heat riser.

If the idea of embedding the barrel in a partition wall is irresistible, be sure to provide expansion jointing (such as non-combustible insulation or braided fiberglass gasket) where the metal meets the masonry, and well-sealed maintenance access to the interior of the bell. Or better yet, resist the idea, and build the partition about 4 inches away from the barrel with room for future maintenance.

ASTM E-1602 5.7 Heat-exchange Channels: These are supposed to be firebrick, soapstone, or other refractory materials laid in fireclay mortar or other refractory mortar.

Clay-based mortars have better longevity and less movement in high-heat applications than products containing Portland cement, and are more forgiving of rough firing-in than most ready-mixed refractory mortars. Clay basically becomes brick if the temperature is hot enough. It is certainly possible to build the entire cob bench using a commercially processed fireclay with either brick grog (ground-up brick aggregate) or masonry sand (such as ¼″-minus rock-crusher fines). In most areas, we find local clay soils are adequate to the purpose.

As for refractory components: Clay chimney-liner sections with approximately the right system CSA have proven to work with rocket mass heaters as described here. Round clay drain pipe has also been used, where available in the right internal diameters. Firebrick channels can be built with some alterations: their rougher surface imposes some drag. If you want to build the channels of firebrick, use 8″ by 8″ channels for an 8″ heater, total channel length not more than 20 to 25 feet, and minimize turns (use cut bricks or clay-sand mortar to smooth the turns as much as possible).

Metal pipes make the process of laying out the heater much easier for most builders. We feel that stovepipe bedded in clay-sand (cob) handles both corrosion and refractory considerations. Many builders have used galvanized ducting, considering it as sacrificial formwork to give the proper shape to the clay-sand mixture which hardens into permanent, refractory channels. Rocket mass heaters built this way have lasted 20+ years with no reported problems.

One possible exception would be damp conditions: underground heating flues in a greenhouse show corrosion much faster, and unfired earthen masonry is not structurally sound while damp. Damp-resistant materials such as ceramic flue lined channels, and an outer casing of unit masonry, will provide better longevity in damp conditions.

ASTM 1602–03 5.9 Chimneys: The guide specifies either low heat chimney as approved by the local jurisdiction, or factory-built residential chimney (UL 103 HT). Weight of masonry chimneys is not to be supported by the heating channels unless expressly designed for the purpose (generally, keeping load-bearing structures separate from fiddly heating structures is good practice, simplifies maintenance and remodels of both).

We typically go for the HT (high-temperature) chimney, as for a woodstove. Mostly because the parts are generally available, and it alleviates some potential concerns in case anyone did hook up a hotter woodstove in future, or prime the chimney using open flame.

For rocket mass heaters with typical channel lengths as described here, and no bypass option, the exhaust temperature is typically 100°F to 150°F at the surface of the chimney as it exits the bench. You could use a low-temperature exhaust system if you can find the parts, and are confident that your installation will not be mistaken for a HT chimney by future owners.

Questionable Changes

Outside Air: (ASTM 5.6.1, some jurisdictions add this to R1002, despite it being omitted from the current version of the codes.)

Some states retain a requirement for outside combustion air for all solid-fueled devices.

This requirement was developed in an attempt to reconcile open fireplaces (which can suck up over 500 cubic feet per minute of combined room and combustion air[1]) with late-20th-century airtight homes. At one point, it was seen as a desirable energy-efficiency goal for a home to be as airtight as possible. Following the diagnosis of "sick building syndrome," we now recognize that both building and occupant health demand a minimum fresh air exchange of about one-third of the home's air volume per hour.

Local building codes are still re-adjusting to these contradictory goals. In some cases, airtight building standards have been combined with a requirement for a minimum size of screened vent directly through the wall. This appears to be regulation getting carried away with itself. Previous best practice standards managed both fresh air and weather-sealing with more art and less fuss.

Outside combustion air is air that comes directly into the firebox from outside the insulated building envelope. (If air is brought into the room instead, it is called "make-up air.")

In practice, there are several problems with outside air as a solution to negative building pressures:

- Outside air must always come from below, never from above the fire. This reduces the risk that the outside air intake will warm up, reverse flow, and begin to act as a second exhaust chimney. This makes it difficult to construct a proper outside air intake for basement or some ground-level installations, or for a down drafting J-style firebox. (Note: Outside air must NEVER be brought into the bottom of a J-style firebox below the feed; this breaks the siphon, allowing the feed to act as a second

chimney, and smoke to escape upward into the room whenever the fuel feed is open.)
- For solid-fuel burning devices which are loaded from inside the room, the fuel-loading door must be opened sometime. Outside air does not prevent smoke from escaping into the room when the firebox is opened to add more fuel, and some builders feel it contributes to turbulence at the fire opening and may even increase smoke escape.
- Outside air intakes on most wood-burning stoves bring air into the firebox, often at a point in front similar to the room air intake from a vented door. In the case of a rocket mass heater's J-style firebox, this point would be just above the fuel opening, violating the rule against delivering outside air from above the fire.
- Outside air intakes are a breach in the building envelope, and must be made impervious to fire. This means they're often made with metal, and can conduct heat outward or cold into the building. Good detailing can address these issues, but expert fireplace builders have been surprised at how much heat travels along this path, in both directions (chilling the room, and/or heating up combustible parts of the building envelope along the path).

Our preference would be to resolve any negative pressure problems in the building in other ways, such as opening a window while burning the stove, or closing communicating doors if it is necessary to run a vent fan (in a bathroom, kitchen, etc.) while firing the heater. Balanced, heat-exchanging ventilation systems are available to ensure that the building has adequate supplies of fresh air without sacrificing much heat in the process, and simple DIY air-to-air heat exchangers can be built for windowsill operation for about $15 in parts (example plans for some heat recovery ventilator designs can be found at builditsolar.com).

Rocket mass heaters do not draw anything like the air volume of a poorly designed fireplace, and they are burned for only a few hours per day. Most of the time, the air intake is shut down while the mass releases stored heat. And of course, the heat stored in dense thermal mass vastly outmatches any heat lost to a little extra fresh air.

If outside air is absolutely required by local officials, it is possible to build an outside air feed onto a rocket mass heater. It would be contrary to the manufacturer's recommendations, if we ever get them listed, but it can sometimes be done in relative safety. See Appendix 3 for one possible design solution.

UL Listing for Rocket Mass Heaters

There is not currently a feasible EPA testing protocol that can be physically conducted on a J-style firebox. Two protocols exist for solid-fueled appliances, one for wood stoves, and one for batch-burn masonry heaters. We anticipate that even if the funding appears within the next year, it will be several years before a testing protocol can be developed and approved, and only at that point could rocket mass heater designs be EPA tested to a consistent standard.

Meanwhile, grassroots designers continue to refine various designs to amazingly low emissions limits. (One prototype built in fall 2014 by Peter van den Berg displayed a 30-minute run at less than 20 ppm of CO, most of it around 6 to 12 PPM — an astoundingly low number in a field where anything under 100 to 500 is considered quite good, and numbers above 5000 are associated with visible smoke. See Appendix 3.)

The beauty of mass heaters, of course, is that their efficiency and clean emissions are effectively multiplied by the hours spent "coasting" compared to actual firing. A heater that operates at 80% efficiency while firing for 4 hours would be over 97% efficient if you consider the full 24-hour heating cycle fed by that 4-hour burn.

Nothing in the current testing protocols allows for this type of quantum leap. It's as if the regulators have resigned themselves to everyone operating their wood burners 24/7 — with the operator literally asleep a third of that time — yet are trying to somehow regulate smoldering overnight fires up to the same clean-emissions standard that can be achieved at peak burn. The operator actively works against it; he wants to wake up to a warm floor in the morning much more than he cares about retaining the stove's clean emissions statistics once it's out the showroom door.

If we take away the assumption that burning fires overnight is necessary, and rediscover the tricks our ancestors developed over millennia of masonry tradition, we may actually achieve warm mornings with smog-free horizons.

As I finish this edition, our county (Okanogan, WA) is under widespread evacuation notice for large-scale wildfires, with air quality ranging from "unhealthy" to "hazardous" due to drifting smoke. The sun and moon, when they can be seen, are the color of copper in the sky. My hair is sticky from the smoke, despite three extra washings this week.

For fire protection, we residents are encouraged to clear small brush and "ladder fuels" from around our homes. The small firebox of the rocket mass heater practically invites the use of this fuel, traditionally considered "women's wood" because it can be gathered by hand, without ax or saw.

The current reality is that rural residents who heat with propane, electricity, or logs cut in the national forest often dispose of their stick fuels by piling and burning them in the brief rains of spring. As a result, there is no season free of airborne smoke in our region. Fall merges wildfire into woodstove season; spring sees woodstoves and burn-piles, and summer is wildfire season.

I could literally clean the air around here by running my stove more, if I could think of anything to do with that much heat.

I hope that the guardians of public safety will see the advantages as well, and that the regulatory environment will increasingly favor this marvelous and efficient option.

Notes

1. The Chimney Sweep's Library: http://www.chimneysweeponline.com/howoodbtu.htm
2. David Darling, Encyclopedia of Alternative Energy: http://www.daviddarling.info/encyclopedia/W/AE_wood_heat_value_BTU.html

Codes and Standards Sources:

3. The European standards descriptions were gleaned from www.stovemason.com/about/, www.techstreet.com, and https://global.ihs.com in August 2015; the authors have not yet purchased full copies.

4. General info from David Lyle's 1984 *Book of Masonry Stoves,* Chelsea Green, 1997; names translated online.
5. http://publicecodes.cyberregs.com, specifically http://publicecodes.cyberregs.com/icod/irc/2012/icod_irc_2012_10_par028.htm
6. ASTM E-1602–03, Standard Guide for Construction of Solid Fuel Burning Masonry Heaters. (The "03" indicates year of initial approval; it was reapproved without changes in 2010.) May be obtained from ASTM International, www.astm.org.
7. Derived from the booklet *A Guide to Residential Wood Heating,* publication 62310 by the Canadian Mortgage and Housing Corporation, 2008, and from online sources such as Installation Clearances for Wood Stoves at www.woodheat.org, www.woodheat.org/clearances.html.

FAQs

How It Works

How can a rocket mass heater use so much less fuel than a woodstove?

There are woodstoves on the market being advertised as "75% efficient" or better. People often ask how it's possible that a rocket heater could use ¼ to ⅒ the fuel. Is it physically possible for something to be 750% efficient?

It turns out that even at their most efficient burn, woodstoves are "graded on a curve" to compensate for the mandatory minimum woodstove chimney temperature of 350°F (to limit creosote deposits). Many woodstove professionals use a lower BTU value for the firewood, about 15% less, to factor that mandatory heat loss. (Theoretical BTU content is about 8700 BTU/lb; a woodstove can be expected to generate about 6200 BTU/lb after factoring in water weight and chimney heat loss; www.chimneysweep online.com.) So a "75% efficient" stove, run perfectly on its lab-testing firing pattern, with wood at 15% moisture, might be expected to deliver 60 to 65% of the wood's theoretical BTU value into the home. The rest of the energy would go up the chimney as smoke, steam, and heat; or remain in the stove as unburned charcoal.

Masonry heaters can burn hotter and cleaner (1200° to 2000°F, which would make metal stoves glow cherry red and void their warranties). With a hotter and cleaner fire, and operators who stick to the instructions because they consistently produce comfort that way, masonry heaters don't run the same risk of creosote deposits and chimney fires. So these heaters can exhaust more efficiently, often at temperatures around 200°F. Then they totally blow the curve by coasting on stored heat.

If an RMH was 85% efficient during operation for 4 hours, then coasted for 20 hours with the air flow completely shut down for near-zero heat loss, could we consider the overall average efficiency as something like 97.5%?

The bigger problem is that woodstoves are rarely operated at peak efficiency once they leave the testing lab. Owners who want overnight heat may use all kinds of tricks to slow down the fire, causing it to smolder inefficiently. A choked-down woodstove may be working at less than 15% efficiency, or even at a negative efficiency (drawing excess warm air from the house toward the end of the fire). It is difficult to achieve steady comfort with a woodstove, but some owners are willing to sacrifice a lot of wood in the effort.

Rocket mass heaters also "cheat" in several other ways. Heat loss calculators designed for forced-air heating or on-demand radiant heaters don't fully explain how people can be comfortable with the tiny amount of fuel a rocket mass heater actually burns. The mass absorbs ambient warmth, maintaining free heating and cooling through the shoulder seasons. Warm seats and cooler walls and ceiling mean less heat loss from the house in every way. And the opportunity to indulge in a "personal sauna" by sticking toes under the bench cushions becomes far more significant to comfort than trying to maintain a theoretical room temperature in the back-bedroom closet.

The overall result is what we see in practice: a family which used to bring in three large tote bins of wood per day (15 cubic feet or more) to feed their certified woodstove can find the same wood storage bin lasting for three days ($1/9$ the fuel consumption) with their new rocket mass heater.

I have a small child/frail elder/pet. How can I keep them from getting burned?

Rocket mass heaters are generally safer than metal woodstoves and fireplaces. Toddlers can't easily see the fire, or reach the hottest parts (upper third of bell, down in the fire itself). Most people and pets have no difficulty staying clear of the hottest parts, and can safely enjoy the warm mass.

If someone in the household is subject to loss of bodily control, falling, intoxication, or memory problems, you may want a physical barrier around the hottest surfaces of the bell. DO NOT cover the barrel with insulation. See "Barrel Decoration," Appendix 6.

Some families with very sensitive individuals have divided the heater into separate rooms, passing the bench through a heat-tolerant wall, for a safe warm area completely separate from the combustion unit. DO NOT place a wall between bell and feed — there simply isn't room to do it while preserving functional draft and maintenance access.

Neither children nor fire should be left unsupervised, especially not together. In case of a serious burn with blisters or broken skin, seek medical advice.

Functions

Will it replace my furnace?

A rocket mass heater is not a furnace nor a boiler. It is a large radiant heater which provides ample heat for human comfort. Like all radiant or contact heaters, is best located centrally in the occupied space, in line-of-sight of any rooms needing heat.

Furnaces and boilers located outside the occupied parts of the house sacrifice efficiency for convenience. They lose heat in transmission. They also require automated controls, such as thermostats, fuel augers, pilot lights, fans, or other draft boosters.

All this automation uses energy, and makes them vulnerable to power outages. But you can't beat the convenience. Some people use both.

See Chapter 2 and Appendix 4 for more details.

How big a heater do I need?

For most homes in cool and cold climates, we consider an 8″ diameter system. 6″ systems may be useful as one-room heaters, or for smaller buildings in mild climates. The storage mass will depend on your space's height, layout, and intended use.

Very large homes may need more than one heater, supplemental heat for occasional rooms, or a larger type of batch-burn masonry heater or industrial boiler. Consider reducing the heating load with insulation, weatherization, and zone heating design.

Can I use it to cook and bake?
Can it supply hot water?

The primary purpose of a rocket mass heater is for home heating and cooling.

Cooking and hot water are needed year-round, but heating is a seasonal need. Would you want to heat the living room as well as the kitchen every time you cook in summer?

We like to have separate cookstoves and heating units, perhaps even an outdoor summer kitchen, but in a pinch you could do a bypass on your rocket mass heater for summer cooking.

A rocket mass heater does provide sufficient stovetop heat for emergency cooking. It's more often a simmer than a full boil, but gets hot enough for soup, tea, and delicate French sauces and desserts. The stovetop itself can also be used as a griddle, with care.

It's a design tradeoff between quick response (useful for fast cooking) vs. stable, stored heat (useful for overnight heating).

We don't recommend water heaters as a DIY project in any case, due to the risk of lethal steam explosions, and the high level of skill and experience needed to plumb hot water safely. A pot or kettle is far easier and safer to manage.

Firebox

How does the fire burn sideways and upside down? Doesn't it smoke?

It's called a thermosiphon.

Picture a water siphon: a hose full of water, one end in a tank, the other end outside and lower than the level in the tank. When the fall is great enough, the weight of the water in the long outside end pulls water up over the rim of the tank to drain the tank. The siphon works as long as there is a difference in height, and the siphon stays full (no air leaks).

Heat rises — heated fluids rise as they are replaced with cooler fluids. So with enough heat, we can turn the siphon idea upside down. A tall, hot, updrafting chimney draws flames or air upward. Cool air flows inward at the bottom. If we connect a very hot, very tall chimney to a downward air intake, we can draw air and even flames sideways and upside down.

Cooled gases no longer rise. So not just height but temperature, and therefore flow volume and length, affect the success of thermosiphon systems.

A leaky thermosiphon doesn't work any better than a leaky water siphon. If you have air holes in the bottom of a rocket firebox, both vertical openings leak smoke upward as you might expect.

The downdraft in the bell requires either cooling, or a warm exit chimney providing onward draft. The insulation around the heat riser helps maintain a difference between rising internal flames and cooling, falling external exhaust. The metal bell sheds heat into the room, cooling the exhaust gases as they fall. Contraflow masonry heaters made of tile or brick don't shed heat this rapidly, so they often use shorter heat-exchange paths and rely on a hotter exit chimney for overall draft.

Why don't my dimensions come out right in the firebox?

Check that your bricks match the given dimensions, and are not irregular. Occasionally you may find that you have inherited a pile of odd-sized brick, and need to adjust the dimensions.

Once you know the interior dimensions are workable, remember to ignore the outside corners. Start by aligning an inside corner, every time. Build a cardboard or wooden form for the inside if you have to, and lay the bricks snug up against it. Work in a spiral, so that each brick has an inside corner butted up against the previous brick. The outside corner can extend out unevenly into space. This gets easier with practice.

How do I make the firebox with a different size of brick?

Make the height as close as possible with some combination of your available brick. Adjust the width to make the proper cross-sectional area: a 50 to 52 square inch opening for our standard 8" heater.

For example, if you have some 4" by 2" bricks that just won't make a 7" tall burn tunnel, you could use two courses the 4" way to make an 8" tall burn tunnel.

- Divide 50 square inches by the new height, 8", to find the new width you need: 6.25 inches.
- The feed opening and heat riser would also be 6.25" wide, so these openings need to be 6.25" by 8".
- The bridge over the burn tunnel is easiest to make with full-size bricks. You may need to adjust the burn tunnel length so there is room for 4, 5, or 6 bricks plus the two openings (22" to 25").
- Adjust your template to any new dimensions, and maintain those dimensions throughout.

Making a heat riser 7" by 7.5", or 6.25" by 8", is not practical with shorter 8" brick. There is no support at the corners. Either use the materials specified (9" long firebrick), or cut the bricks into shorter lengths so you can use 2 trimmed bricks for the long sides.

Can I make the firebox a lot bigger? Can I make a miniature for my camper?

Scale changes really mess with fire. We have yet to see a 4" mass heater that works well, although this size can work for smaller radiant heaters or cookers. Smaller stoves may be smokier or require more tending than a larger model, just due to the variable nature of wood fuel and more delicate draft balance. We would tend to build a 6" system for spaces that can take the weight, and just run it less.

The few examples of 10" and 12" mass heaters we've seen are scary-hot. Outdoor prototyping is definitely recommended: a

30% increase in diameter more than doubles the heat output (and fuel consumption, and strains the heat tolerance of most available materials).

If you really want to burn a lot of wood at once, you may want a batch-burn masonry heater, or a private consultation for light industry designs (such as wood-drying kilns).

Can I make the feed taller to take longer firewood (18", 24", 48")?

Because the thermal siphon depends on a difference in height, making the feed taller requires making the heat riser much taller yet. The heat riser must be three times taller than the feed, so a 24 inch feed would call for a 72" (6-foot-tall) heat riser.

This added height sometimes just doesn't fit in rooms, especially if you want to be able to lift the barrel over the heat riser for maintenance access.

It's probably simpler to cut firewood to standard lengths of about 15–16". Shorter cuts dry faster, too. To limit saw time, consider collecting smaller branch-type fuels normally discarded as yard debris.

Can I make the fuel feed bigger, horizontal, or more like my woodstove?

Most woodstoves produce significant smoke during routine use, which can turn into creosote. Therefore by law, woodstoves have a minimum draft temperature of 350°F to keep smoke from depositing creosote in the chimney. To safely operate a heater which extracts more warmth from the exhaust through a heat-exchange mass, you MUST have a clean-burning firebox. In short, if woodstoves are what you know, you might need to un-learn some things to operate a clean-burning heater.

There are promising batch-box designs in development, but they require precise air supplies and fire layout to operate cleanly. Some promising larger heaters are reviewed in Appendix 3. We have also seen numerous heaters that don't burn hot or clean enough, or where a disproportionately large firebox leads to smoke that can leak, backflash, or deposit creosote.

If you know you will need a much larger heater for your climate or home, consider proven types of batch-burn masonry heaters.

Thermal Mass

How do they start when cold?

All chimneys draw better when they are significantly warmer than outside air. As long as the heater is operated at least a few times per week, the system remains warmer than outside air and is easy to start. Coping methods when the mass is colder than outside air are described under "Cold Starts," Chapter 5, Operations and Maintenance.

In warmer climates, where the heater may be used on mild days or rare occasions, we often design a shorter heat exchanger and hotter exhaust.

Doesn't all that mass take a long time to heat up?

The lag time is proportional to your heat storage, and unavoidable. If we are not using it regularly, our heater may take a day to warm up from 55° to 75°F or so, and two days to cool back down. The radiant heat from the metal bell is instant, however, and offers an immediate sort of comfort while the mass warms up.

If this sounds inconvenient, consider other forms of wood heat. A responsive wood-burning stove has a tendency to over heat easily and cool off quickly. Boilers can be run at a relatively steady temperature, but require regular outdoor feedings of enormous amounts of wood (due to losses in transmission in all cases, and smoky, inefficient fires in many designs).

Most people want a stable, comfortable indoor temperature. Thermal mass excels at stabilizing the temperature, in all seasons. It usually keeps us more comfortable with less work; we learn to plan around the lag time.

To save space, can I make it taller, or build it into a wall?

It is certainly possible to make the entire mass tall (like a masonry wall), or compact (like a cube). Another option is to sink the mass into the floor, so that only the combustion unit sticks up into the room.

Some European masonry heaters are built as dividing walls between 4 or 5 rooms. There's no reason you couldn't do the same with a rocket heater, if you have the patience to work around local code requirements. (See Chapter 6.)

No heater should be part of a foundation or external wall; the heat lost to outdoors and to ground damp can be enormous.

Can I fit it behind my current hide-a-bed sofa?

You won't get the full benefit of the heater if you tuck it out of the way where you can't sit directly on it. They are designed to heat people, not space.

Think of the heated bench as a luxury you can offer your guests: If you wanted your entire family and guests to sit and relax on your new heater, where would you put it?

We recommend displacing your favorite sofa and putting the heater where it will get the most use, as permanent built-in furniture. A heated bench, bed, sofa, floor, or workbench lets you continue to use the space the heater takes up. Most owners will get more satisfaction from the heater, both financially and physically, when they sit directly on it.

That's a lot of weight. Can I make a heater like this, but with less mass?

"Lightweight" heaters, by definition, cannot store heat in thermal mass. The old minimum definition of 1500 lbs would be a very small (single-room) masonry heater.

There has been some interest in "granular" mass heaters — big boxes or concrete bins filled with gravel and rocks. However, these heaters can be even heavier than solid mass versions, while storing less heat and using more fuel. For now, "portable" mass heaters are relegated to Appendix 3.

Our current recommendation for spaces that can't be altered or take the weight, such as campers or rental homes, is to improve insulation first (drapes, wall art if you can't alter the exterior walls). Thermal mass can sometimes be added within the safe limits of the structure's design, such as masonry splashbacks or surrounds behind heat sources.

Materials and Methods

Do I have to make it out of dirt? Why do you make it out of dirt and not concrete?

As it turns out, dirt is more comfortable to sit on than concrete. Also, clay-rich materials holds almost twice as much heat per pound as a typical concrete. But you can make a concrete box if you want the look and feel of concrete in your home, and are willing to sacrifice efficiency and comfort.

See Chapter 4 and Appendix 1 for more on this topic.

Can I make a metal firebox?

Metal is not suitable for high-temperature fireboxes; it will warp and burn out with extended use.

What else can I use for ... (insulation, firebrick, barrel, etc.)?

See alternatives in Chapter 4 and Appendices 1 and 6.

Site/Location

Can I put a rocket mass heater in my basement?

Not recommended unless it's part of the lived-in home. It's best to locate a mass heater in the occupied parts of the house, where it's efficient and convenient to operate.

If your (daylight) basement is where you spend most of the day or evening, tending a fire there might be convenient. But most basements are occasional-use spaces at best. The differences between a basement and ground-floor project include:

- Uninsulated cellar walls soak up a lot of heat. Locating this type of heater in an unoccupied basement adds a massive, poorly-insulated space to the heating load.
- Basements may present severe problems of ventilation, damp, and negative air pressure. The lowest parts of a building naturally draw in cool air, and may have negative pressure problems. Basement heaters will definitely need a full-height chimney to above the roof ridge; do not attempt to vent a basement stove out at ground level.
- A seldom-used heater does not save fuel, and is not much good for emergencies, as it's difficult to start a fire with a cold chimney.

Better heating options may include:

- Build up a footing to bring the heater into the occupied space, or flush with the floor.
- Locate the heater on an existing slab such as an attached garage, or in a daylight basement, convenient to everyday activities.
- Build an addition off the main living areas of the house for easier foundation work, and insulate it well.
- Consider a different type of energy-efficient heater that is designed for basement installation and automated operation, such as a decent furnace or boiler.

Can I build it over my existing floor?

We have seen a few rocket mass heaters built over existing, suspended-wood floors.

We prefer to see heaters built on noncombustible foundations. This has been standard practice for fireplaces and masonry heaters for a very long time — long before the building codes made it a legal requirement.

If your floor happens to be an existing concrete slab, that's great luck. A 4″ concrete slab on grade will easily hold the weight of a low masonry bench, up to about 30″ average height (or an 18″ bench with up to 48″ seat back). Even compacted gravel (as for many greenhouse or barn pads) can work well, with damp protection.

The problems with building heavy masonry over a suspended wood floor include:

- Weight support: Most wood floors are designed for dead loads of 40 to 100 pounds per square foot, and a rocket heater typically brings 130 to 225 lbs/sq. ft. for a low bench (more like library stacks or a large aquarium).
- Movement: Wood tends to flex with time and weather; masonry doesn't flex well, and needs stiff supports to prevent cracking.
- Fire danger: Raised air gaps and a lot of extra insulation on the underside of the heater can protect the floor — as long as they remain intact. The air gaps must be big enough for good flow, monitoring, and cleaning; operator neglect could increase the hazard.
- Moisture compatibility: Masonry can trap moisture or condensation against wood, accelerating rot.
- Finally, the extra materials and air-channels can make the heater a bit tall for comfortable seating and cooking, and add more weight.

We definitely prefer noncombustible footings, and most building codes require them.

How close can the heater be to nearby walls, windows, or walkways?

Minimum clearances and thicknesses are heavily regulated. Your local building office may insist on particular specs to meet local code. See Chapter 6.

How do I deal with a combustible wall or load-bearing post too close to the heater?

One engineer suggested replacing a 4x4" wooden post with a 4" diameter metal post. In some areas "rock jacks" are popular — a metal post filled with small-aggregate concrete.

Include an expansion joint or thin air-gap between metal and masonry, so that their different movements will not cause cracking.

How can I have the heater closer to the wall, or go through the wall?
Can I locate the bell and firebox in one room, but have the bench go through the wall into another room?

See the discussion of wing walls toward the end of Chapter 6.

Can I locate the bell and bench in my room, and just have the feed outside?

We do not recommend trying to locate a wall between the wood feed and the barrel. There simply isn't room while retaining the proper proportions. Many versions we've seen are a serious fire hazard, and even the most prudent versions tend to make the stove cantankerous to operate and maintain.

Can I use my existing masonry chimney?

If you have a masonry chimney in good condition, with a suitable flue size, you can use it for a rocket mass heater. Exposed masonry chimneys (on the outside of the wall) should have an insulated liner, insulated chase, and/or a way to preheat the chimney when it is colder than the house.

Flue size should be approximately the same cross-sectional area as system size— see Chapter 6 for compatible flue sizes. Oversized chimneys should be lined with an insulated liner to match the new heater.

Safety and Code: Clearances, Foundations, Thickness

Can I just use the design you present in this book?

With our compliments. However, we can't guarantee that anyone else will approve it. Please look into your local jurisdiction's applicable rules, and make informed decisions about compliance.

Do I really need to ... (meet all those other specifications)?

For your convenience, we have removed over 50,000 superfluous or unnecessary words from this book. What remains is the information we consider critical for reliable results.

Questions about cutting corners are a matter of function and risk. "What am I risking if I don't do X?" The answer depends on what you do instead. See the "Warnings" section of Chapter 5.

We have laid out the workable alternatives we personally would still consider doing, having tried many others. We've included options that may not meet current codes, because codes are not the same everywhere and they do change with time.

If the risks seem tolerable and you want to experiment, do your own research, and build in the option to fix it later.

Do I really need a vertical, through-roof chimney?

(That "other book" says these heaters can exhaust out the wall like a dryer vent.)

The first books about rocket mass heaters were written by a man whose house is 100 "round feet" and has no roof vents. Most of the other proponents are his students. Unfortunately, unconventional sideways "chimneys" do not work well in conventional homes.

We originally covered this issue, in detail, in almost every chapter. We've winnowed it down to just the even chapters. See "Chimney Methods: Up or Sideways?" in Chapter 4, and discussions in Chapters 2, Chapter 6, and Appendix 3.

Can I use the same (shared) chimney flue for the rocket stove and another appliance?

Not safely. Exhaust from one appliance can backdraft down the other. It's better than no chimney, but not by much. It's not physically possible for the flue to be the correct size for each device operating alone, yet also the correct size for both devices operating at the same time.

With a relatively low-temperature exhaust like a rocket stove, the exhaust can easily leak back down into the house through the other device. Separate flues are the best approach, and the only one allowed under code in most areas.

Do I need a masonry footing or foundation?

It's a good idea, and may be required by your local building codes. This is hard to fix later. See discussion in Chapters 2 and 6.

Can I reduce these clearances?
Do I really need a 4" air gap behind the "zero clearance" heated bench?

The bottom line is that any nearby combustibles, such as wood trim, wallboard, or cushions, must stay below 165°F for safety (75°C). Hazardous temperatures begin as low as 185°F / 85°C (plastics, paints, and wood finishes begin to degrade, and may become more flammable with time). Anything combustible that is too hot to touch is a fire hazard, and needs to be removed or properly heat shielded before you can safely use the heater.

It's generally easier to build in an extra inch of clearance from the beginning, than to move the heater an inch after it's built.

Why does the bench need to be 4" or 5" thick?
Can I make it thinner (to reduce mass, size, or for quicker heat-up and cool-down)?

We have seen problems with both hot spots and structural integrity in benches where the masonry was thin. A cob surface of 2" or 3" can create a hot spot over the pipes, sometimes hot enough to damage synthetic fabric. This can be fixed by adding more masonry or plaster atop the bench.

It is harder to fix structural problems when the pipes are crushed due to insufficient side-wall support. Often, the heater must be torn apart to replace the crushed pipes. This problem is most common during construction: someone stands on the wet bench, or tries to build on top of the ducting before the sides are firmly supported.

Aesthetic Concerns

Do I have to put an oil drum in my living room?
Can I replace or cover the barrel with something more attractive?

It doesn't have to look like an oil drum, but it should work like one. A weathered or blackened steel cylinder gives great radiant heat distribution. Whatever bell you install must be about the same height and width. If using a masonry bell, the radiant heat and downdraft properties will be different, and therefore the draft proportions may need to be adjusted.

There are ways to decorate the barrel, or upgrade it, without compromising function. See the "Barrel Decoration" section in Appendix 6.

One dilemma in photographing attractive rocket mass heaters is that the quality of heat from an exposed barrel feels so nice that owners learn to ignore the looks. We removed a decorative copper sheathing from our Cabin 8" heater because we preferred the quality of heat from the steel.

Can I use cushions or a mattress on top of the heated bench?

When built to the design standards described in Chapter 4, the bench can have heat-tolerant cushions or blankets in contact with the surface. See Chapter 5 for suitable fabric ideas.

Do I have to make it out of mud?

No. See "Materials and Methods" above, and Appendix 1. Earthen-cored heaters can be finished with natural plasters such as clay, lime, gypsum, or with stone or tile details. For a totally brick or stone facade, it's best to build with these materials and then fill with earth or clay-based mortar.

Performance and Troubleshooting

Is there someone who can build one for me?

See "Resources" in Chapter 5 for good places to look for builders or advice.

Expect a professional installation to cost more — a realistic bid will have at least four figures. As with any remodeling job, "unforseen problems" are common, and experienced contractors bid with this in mind.

Common Errors by Location

Firebox Errors

Common Error: "Outside Air"

Introducing "outside air" to the firebox anywhere below the feed opening causes smoke-back from the fuel-feed opening. Extra air breaks the siphon, and the feed becomes a second chimney. If outside air is needed, it must enter through the feed opening (see Chapter 6, Appendix 3).

Common Error: Metal Firebox

Your typical barbecue or cook-stove does not get this hot. At the temperatures routinely achieved in a clean-burning rocket firebox (1200–2000°F), metals warp and expand, can crack any surrounding masonry, and most will oxidize and become more brittle within the first heating season. Metals are not suitable for a rocket mass heater firebox. Firebox interiors should be made of suitable refractory materials as described in Chapter 4 and Appendix 1.

Common Error: Incorrect Firebox Proportions

The firebox acts as a unit. The heat riser must be tall, the fuel feed short, the burn tunnel just long enough but not too long, and the cross-sectional area (CSA) consistent through all parts. If the heat riser is reduced in height, CSA, or insulation value, or the fuel feed or burn tunnel are enlarged, the firebox does not work properly.

Fuel Feed Errors

Common Error: Sideways Fuel Feed

Many people have tried converting the vertical J-tube to a horizontal or diagonal L-feed, some with more success than others.

The trick with a horizontal feed is to maintain a perfectly clean burn. Without attention, a horizontal fire may smother itself, causing smoke and creosote. Smaller fires tend to go out; larger fires make a large volume of smoke that needs somewhere to go. Adding a door offers air and smoke control on the room side, but may just exhaust smoke outward and worsen the creosote problem.

A diagonal feed offers the worst of both worlds: fuel does not self-feed and tends to smother itself, yet there is a large warm space above the fuel that can draw smoke backwards into the room.

If the J-style firebox featured here does not meet your approval, we recommend looking into other well-developed heater designs. The Holmes/van den Berg batch box rocket heaters show promise as a horizontally-fed system, with different heat-exchange proportions and rules of thumb. (In our opinion they require more discipline to run cleanly, but when operated carefully they give excellent results.) Many traditional masonry heaters are also designed for batch loading. The J-style firebox was designed to minimize fuel use, not to replicate familiar large-volume fireboxes.

Heat Riser Errors

Common Errors: Heat Riser Not Insulated

Poor insulation or air leaks between heat riser and bell can interfere with downward draft in the bell. If the hot exhaust stagnates at the top of the bell, it can drown the fire in its own exhaust. Poor insulation will often display a delayed reaction, where the heater drafts fine at first, then the draft suddenly fails about 30 minutes later (as the heat riser material heats through). Stagnation can result in smoke pouring out the feed. Fix before using the heater again.

Common Error: Heat Riser Too Short

Some builders who have only seen the firebox operate as a stand-alone outdoor system (such as a rocket cookstoves or camp stove) underestimate the necessary height of the heat riser. The heat riser is the primary engine for the system, drawing flames upside down and sideways, and providing the initial push to get things started through the heat-exchange channels. The heat riser should be at least three times as tall as the fuel feed, and at least twice as tall as the burn tunnel, as shown in Chapters 4 and 6.

Bell/Barrel Errors

Common Error: Covering the Bell

Many people don't love the appearance of a weathered steel cylinder in the living room, so they experiment with ways to cover it. Any insulation in contact with the bell can cause the system not to draft properly — similar to the stagnation problems experienced with an noninsulated heat riser, above. Up to one third of the metal surface can be covered with dense, noncombustible decorative materials without affecting performance: earthen masonry, tile, plasters, or similar materials. Heat shielding suspended on or behind the bell can also be used to mask appearance, converting some radiant heat into upward-moving warm air. See "Barrel Deco Design Challenge," toward the end of Appendix 6.

The bell should never be embedded in a load-bearing wall or other structures — this makes repairs or maintenance access extremely difficult, and the heat expansion of the hot metal at the top of the barrel often cracks any masonry in contact.

Common Error: Unsafe Clearance from Bell to Combustibles

A bell that does not have 5" thick masonry sides would be required to have 36" clearance under IRC section 1002 for masonry heaters. A non-certified (metal) woodstove under most local codes requires a similar clearance, but this clearance can be reduced with heat shielding. A standard, wall-mounted metal heat shield with 1" air gap, and 3" air gap top and bottom, would allow the metal stove to be 18" from the wall. The greatest permissible reductions that we have seen for non-certified woodstoves would allow a minimum 12" clearance with very good, multi-layered heat shielding.

If a combustible wall gets hot enough, more than about 150° to 160°F, the wood and paper slowly bake and become more flammable over time. Any signs of discoloration or other changes (melting, warping) in combustible materials, or any combustible materials that become too hot to comfortably touch during heater operation, means the clearances are not adequate for safety. Remove combustible materials and/or improve the heat shielding before using the heater again.

Common Error: Air Leaks in Manifold or Bell

Even if no smoke escapes, if air can get into the bell, the exhaust can stagnate at the top of the bell. This can cause the fire to smother in its own smoke, and to smoke back into the room. Make sure the bell and manifold are well sealed, including any nearby cleanout access door(s).

Manifold Errors

The manifold is difficult for people to envision and construct.

Common Error: Manifold Not Big Enough

When transitioning from the side of the bell into the horizontal heat-exchange bench, it is easy to leave a too-small slot. Peter van den Berg suggests that if it's a right-angle turn, the opening should be at least 150% of the systems' cross-sectional area (CSA) to avoid flow restrictions. If the pipe rests on the bottom of the manifold chamber, then the bottom of the pipe receives no air flow (it all comes from the top and sides), and should not be counted toward the total flow area. For slot-like openings with a pipe in the side, both Peter and Erica feel that the minimum slot width is about half the pipe's radius: 4" for an 8" system, 3" for a 6" system. Ernie has demonstrated that some systems work despite slight restrictions (down to about 2"), but all builders agree that smaller manifolds can more easily become clogged with ash that impedes the heater's draft. If a heater that has operated reliably through most of the season starts having draft problems, clean the manifold.

Heat-exchange Errors
Common Error: Heat-exchange Channels Improperly Sized

There are currently two models for how to size the heat-exchange passages: the Evans channel system, and the bell system. The channel system requires consistent CSA throughout, smooth turns, and no unnecessary roughness or corrugations. Stovepipe, steel ducting, or clay chimney liner or drain tiles have all been found to work well. Do not use corrugated material unless nothing else is available (see below).

The bell system, used by builders such as Peter van den Berg and Matt Walker Remine, has a completely different set of rules. A bell is a heat-trapping cavity that is at least four times the system flow area, and has its entrance and exit quite a bit lower than its top. If the entrance and exit are near each other, sometimes a short wall or baffle is provided to stop exhaust flowing directly from one to the other. The goal is for the hottest exhaust gases to rise, stratify, and give up their heat at the top of the bell, while only the cooler exhaust gases escape from the low exit opening and proceed toward the chimney. "Cooler" is relative, as the exhaust gases must still have enough heat to rise up the exit chimney. Bells are commonly sized to the system's heat output by comparing their internal surface area; metal and masonry have slightly different heat-extraction rates.

If the heat-exchange channels vary in size, for example by using the wrong size elbow or transitioning back and forth between two or three available sizes of pipe, it creates a lot of drag and can cause the system not to draft adequately.

Common Error: U-trap or Downward-flowing Exhaust Structures

Horizontal pipes of consistent CSA, with a slight rise toward the exhaust, work well in the heat-exchange bench. Pipes that go up and down create heat traps: hot exhaust stagnates at the top, cooler exhaust and condensates (water or fog) get trapped in the low spots. If you need the exhaust to enter and exit at very different elevations, consider a bell (four times the system area) to avoid U-trap-type restrictions.

Common Error: Corrugated or Rough Channel Liners

If using the linear channel version, any roughness creates drag which can slow the system down or even effectively plug it with its own exhaust. It is common to want to use corrugated, flexible ducting such as is often used with forced-air systems or mechanically vented exhausts. But most builders don't realize that each foot of corrugated material creates the same turbulence and drag as about 10 feet of smooth round pipe.

Corrugated ducting, rough bricks, or other rough surfaces create unnecessary drag, and if used the system should be shortened to allow much hotter exhaust up the vertical exit chimney to draw more strongly and overcome the drag. Firebrick channels can be used up to about 20 feet in length.

It is possible to do very short systems with corrugated material, or a shortened system with a short length of corrugated material used for an elbow or odd turn. But it's difficult to successfully estimate the balance between the extreme drag, and the temperature and weather-related draft to run such a system efficiently. Better to cut stovepipe or use adjustable elbows to achieve the necessary turns more smoothly.

Common Error: No Thermal Mass, "Light-weight Mass"

Without thermal mass storage, the efficiency of a rocket heater is greatly reduced. If you do not need stored heat for overnight comfort, but only a quick burst of heat during brief activities like cooking, entertaining, or garage hobbies, consider a purpose-built radiant heater instead of a mass heater.

Heat storage is calculated by weight and temperature of the mass — less mass simply cannot store as much heat. Versions like sand or pebble-style removable mass, with a lot of trapped air, can prevent heat transfer; thus you have unnecessary mass that isn't storing much (if any) heat.

In a portable structure, we can see the attraction of removable mass such as oil or water, that can be drained for transport, but consider the problems of leaks, corrosion, pressure, and steam explosions. Good insulation is lighter-weight, and cheaper to transport, than a large heat-storage mass. One compromise option is to heat with a smaller, conventional cook stove, and put some thermal mass in the heat shielding around it and a big pot of water on top.

Some people consider using just the heat-exchange ducting of the rocket heater with no mass, like a big radiator. However, as the Victorians discovered, creosote readily condenses in such exposed pipes, and a creosote fire's intense heat can twist, warp, or burst open the pipes, spraying burning tar all over the room like a fire-breathing snake. There's also the more prosaic danger of invisible exhaust leaks.

Theoretically the rocket firebox should burn most all the smoke, and prevent creosote problems.

In practice, we feel much safer with the exhaust channels double-sealed in liners and monolithic earthen masonry, and with a surface that allows us to detect and repair any cracks promptly. If using any system other than monolithic masonry (such as a pebble-style experimental stove, or back-filling with loose dirt or gravel), take steps to ensure that the pipe is double-sealed, and have functioning CO and smoke detectors nearby.

Appendix 1:
Earthen Building

Why Earthen Masonry Heaters?

IF YOU HAVE LIVED IN MODULAR-MATERIAL housing all your life, you might wonder why we prefer clay-based, or "earthen," masonry. Do "primitive mud huts" come to mind? Why would we *not* use modern materials like concrete?

It's certainly possible to build a masonry heater using modern materials — including concrete mortars in the exterior casing masonry. But clay is still a key ingredient in the firebox area, for everything from firebrick and cast-ceramic parts, to the mortar (whether it's a traditional clay-based mortar, or modern refractory mortar with clay as just one ingredient).

Masonry heaters are built twice: there's an inner core of the high-temperature firebox and flue channels, and then a separate layer of outer masonry to store heat and please the eye. Because these two layers must be kept separate to prevent cracking (the inner one gets much hotter and expands more), it takes considerable masonry skill to build these layers separate enough yet close enough to work well. If you wanted to go out and buy the components for these heaters, a kit for just the inner core might cost a few thousand dollars, up to five or six figures for a master mason to build a custom installation. These heaters are beautiful and efficient, and most will last several lifetimes, but the cost is daunting.

Were masonry heater builders more common in the past, or was this skill always rare and costly? Masonry heaters thousands of years old can be found in many parts of Eurasia, and even common household versions often last for centuries. With such durable products, even a few master masons per generation can leave a significant legacy. But there is also some evidence that common or "vernacular" builders made their share of heaters, from the fox stoves of ancient Kashmir to the improvised Crimean stoves heating hospital tents during the later days of the British Empire.

Traditional Eurasian masonry heaters were made with the same local materials as ovens, kilns, and fireplaces: stone, clay, and brick. Clay was often a preferred material for sealing the heaters, especially in the firebox area and for big surfaces (floors, large ovens). In areas with suitable earth, literally anyone could go to the "oven earth" hillside, scoop out a few buckets of dirt, and build an oven that would bake good bread for years and years.

Modern masonry heaters may use the full range of available materials, but they still rely on brick and ceramic components for the firebox itself.

What is ceramic made from? Fired clay. Where do we find clay? In the earth.

With the right proportions of aggregates, clay, and reinforcing fiber, earthen building materials offer great performance and durability. As Ianto Evans describes in *The Hand-Sculpted House:* "It won't burn, bugs won't eat it, and it's dirt cheap."

Earth:

- is sturdy
- is plentiful and cheap
- is more comfortable and resilient than concrete or stone
- is easier to shape than wood, stone, and tile
- has low embodied energy
- is nontoxic
- is easily maintained, repaired, or altered
- naturally regulates temperature and humidity without external power

Well-built earthen cottages have performed well in earthquakes and seismic-table testing, although proper detailing is essential (as with any construction method).

Heat does not harm most earthen materials (as it does concrete or lime mortars), and can even strengthen them under the right conditions.

Concrete makes good basements, but earthen masonry makes a healthy house.

Traditional vs. Modern Refractories

Refractory materials are the category of materials used in very high-heat operations, like smelting, glassblowing, foundries, and our clean-burning masonry heater fireboxes. Since we may be reaching temperatures hotter than lava (1800°F), it stands to reason that we need special materials that won't melt or disintegrate while holding our clean-burning fire at these temperatures.

We confess a preference for natural, minimally processed, local materials. Reasons include accessibility, low toxicity, proven performance, sustainability, lower embodied energy, and cost.

Historic masonry heaters were usually made with clay and ceramic clay products. As David Lyle put it on page 67 of his excellent work, *The Book of Masonry Stoves,* " The basic building material in all traditional masonry stoves is clay: clay brick, clay tile, clay mortar. The reason is simple: the expansion/contraction factor. Where all stove parts are made of the same kind of clay, they will expand and contract in unison. This is critical in designing and building masonry stoves. ... In Europe experienced masons will go out of their way to make clay mortar from the same type of clay used to make brick or tile for the stove."

Modern builders may point out that clay does not give as strong a bond as modern refractories. However, we see this as an asset in certain types of masonry work. Softer mortars protect the masonry units, allow re-pointing or repairs, and they offer predictable and noncritical relief for any stress cracking. Too-strong mortars may cause brittle failure; the whole project fails at once, and the masonry units are damaged beyond repair. Builders understand that no installation is truly permanent. Good design allows for future changes, repairs, or even removing and rebuilding the structure somewhere else. A soft, resilient mortar protects expensive cast parts both during normal use and future work.

Clay-based surface plasters or floor layers make it easy to spot and repair any cracks using simple tools like water and a spoon or a pointing tool. Asian k'ang and ondol masonry heaters and Roman hypocausts are examples where the heater was built as a floor or floor platform, often using clay layers to achieve an economical, airtight seal. David Lyle quotes Vitruvius's description, "These pillars should be two feet in height, laid with clay mixed with hair, and covered on top with the two-foot tiles which support the floor." (p.76, ibid.) Clay mixed with hair is a simple description of an earthen plaster or resilient moderate-heat mortar.

Clay-based traditional masonry is a physical bond only: it is soft when wet, hard when dry. Unless it has been "fired" into a permanent ceramic, clay masonry can be resoftened and reworked with water. This makes clay-based heaters vulnerable to water damage, like many other building materials (wood, drywall, and other common materials also require protection from damp), but it also makes repairs easier. Clay is generally nontoxic and safe for children and pets; just avoid breathing the dust if working with dry material (wet it to prevent dust while working).

Modern refractory cements may be harder, form a stronger bond, and come rated for specific heat tolerances. Clean fire burns at least 1200°F in ordinary fireplaces (650°C); rocket mass heaters typically burn hotter, so a material rated for 2100 to 2700°F is needed (1200 to 1500°C). Portland cement and lime products degrade around 600 to 1200°F (below 650°C), so they are generally unsuitable for use in clean-heat fireboxes.

Under proper curing conditions, refractory cements form a permanent chemical bond. Most kinds resist weathering, although some older kinds require rain protection. Remarkable things can be done with modern cements, and their composition varies with intended use. Follow manufacturer instructions regarding preparation, curing time, and personal protection such as gloves, respirators, or eyewear.

Strength and permanence sound like good qualities to a mason — but in the case of stronger mortars, they come at some cost. Permanence and fixed working times require more practice and precision from builders; most novices will spoil a lot of material as they learn. Curing conditions must be met; badly cured material can't be reworked. Strong, rigid materials must be protected from cracking with expansion joints. Cracks will run through the brick rather than just the mortar, potentially causing more structural damage. Repairs may require specialized tools and replacement parts, and these are not always locally available.

Some jurisdictions mandate the use of modern cements or reinforcement. Reasons vary, and may not be fully explained (from seismic concerns, to industry lobbying, to anti-novelty bias). We prefer to meet these safety standards through practical design solutions (such as building heaters that can't fall on anyone, and building them under a roof).

Traditional materials with proper detailing can serve for centuries. Proper detailing for a given climate and material can be verified from historic restoration and trade's expertise. The proper detailing to prevent unforeseen problems with modern materials won't be fully known until they have stood the test of time. To protect your investment of time and labor, we recommend using proven materials where possible.

If local conditions don't require the use of chemical cements, we don't use them. The proven track record of traditional clay-based materials indicates they can last for centuries with good maintenance, or until the roof falls in. That seems permanent enough to us.

Modern Alternatives

Portland cement is not an acceptable refractory material for the firebox. Portland cement (used in everyday concrete) does not handle high heats; it turns back to powder around 600–1000°F. It's not a great material for comfort or health, either: it is toxic and caustic to work with, has a limited working time, is difficult to clean off tools, and can't be re-used once it starts to set. After curing, its hardness makes it uncomfortable to sit or stand on, and tough to remodel or repair.

Portland cement is, however, an ideal masonry material for exposed, damp conditions, and for certain kinds of structural spans, which is why it has become the standard material for foundations, bridge footings, highway ramps, and parking garages. Because it will handle conditions where traditional masonry fails, modern building professionals love it, and sometimes overestimate its appropriate range of uses.

Portland cements can be used in the outer cladding for the masonry heater (for example as cement-based mortars) provided that there is an expansion joint between the inner, clay layers and the outer, cement-type materials.

If any local "expert" suggests Portland cement in the firebox, refuse — with demonstrations, if necessary. (A mock-up firebox built of concrete pavers and suitable insulation, fired to full heat for about a week of daily heating cycles, proves the point pretty well.)

Refractory Cements

If your local jurisdiction won't allow the use of traditional clays, the firebox can be set with a thin-set refractory mortar rated for temperatures 2400°F or higher. (Actual firebox temperatures vary.)

There are several types of refractory cement, each with its own chemical formulation and usages. Look for suppliers or other local users who may build brick ovens, pottery kilns, or industrial refineries. Most refractory cements form a strong, irreversible chemical bond. A few are vulnerable to water damage after curing, and many require special curing procedures.

Modern cements require more attention to all dimensions, including time, because errors are harder to fix. We usually dry-fit the entire firebox without mortar, double-check clearances, and draw a template before re-laying with mortar. For those unused to working with brick or this particular firebox, you may find it worthwhile to make a three-dimensional form out of wood or cardboard, in the exact dimensions of the firebox interior, which you can set the bricks around (after the mortar cures, the form will need to be removed or burned out, so plan its construction accordingly). Consider dry-fitting (without mortar, or with only clay slip) any parts that might need eventual replacement (such as the feed bricks and the first bridge brick).

Not all refractory products form a reliable seal (they bond hard, but we've demolished fireplaces where the cements shrank and cracked in between the bonded portions). Make sure to create secondary seals with high-temperature refractory insulation and outer masonry, so that any air or smoke leaks will not be able to penetrate through all the layers.

Follow any instructions on the refractory cement as far as cure-in and first firing protocols. Some refractory products require damp-cure followed by a slow drying and a warm-up procedure to reach their best strength.

Refractory products that have been used with some success include Cast-o-lite, Sparlite, Cement Fondu, and several others. Most castable refractories eventually crack, and we can all recognize that a small, high-temperature firebox has a stressful job. Experienced masonry heater builders generally design their fireboxes to be rebuilt sooner or later.

Building with Earth: The Oregon Cob Method

Cob is an Old English word meaning "loaf" or "lump." (There are no corncobs involved.) Cob is built with wet lumps of earth, without formwork. (Adobe is earthen bricks; and rammed earth, or pisé, is damp earth tamped into forms.) Cob (or *clobbe, clom, clay daub, monolithic adobe*) was once widely used in Western Europe and other humid climates worldwide.

Cob was the original material used for rocket mass heaters, and it works well. It seals well, is plastic and easy to shape, and is one of the least toxic and least costly forms of masonry. The Oregon Cob method is the name given to the artisan process taught by Cob Cottage Company (of Cottage Grove and later Coquille, Oregon), and described in the publications *The Hand-Sculpted House* and *The Cobber's Companion*.

The Oregon Cob method is sometimes looked askance by established traditional builders, as "fussy" or "labor-intensive." But its batch-by-batch mixing, testing, and amendment process allows science-minded post-industrial folks like us to wrap our heads around the textures that make excellent earthen building, and to gain confidence in making real structures from this under-appreciated material.

In regions where earthen building remains a living tradition, it's far more common to target a handy pit of good-enough dirt, and do whatever's necessary to avoid fiddling with it: make adobe blocks, throw it on the wall with a pitchfork, make the walls three feet thick if necessary. The Oregon Cob method standard wall is 8″ to 12″ thick, and the Myrtle Library in Coquille, Oregon, includes a free-standing curved wall 2″ to 3″ thick (and 3 feet tall) that routinely gets sat on by guests of all sizes. The Oregon Cob standard is more than sufficient, and probably not necessary, for an 18″ tall by 30″ wide mass heater, even with its 5″ side walls. But it makes a pretty good proof-of-concept for the skeptical.

Sculpting a comfortable bench or bed with cob is simple and intuitive. Repairs and remodels can be made by soaking and removing unwanted material, as described in Chapter 5, then remixing and reapplying the same material in the new location. (Try that with concrete!)

Properly detailed earthen masonry is hard-wearing and durable enough to last for centuries, yet can be owner-maintained, remodeled, or repaired with minimal tools and waste.

It's most effective to learn earthen building at a local building project. It is possible to learn on your own, however, especially if you have some prior building experience or other relevant skills. Masons, geologists, potters, bakers, trackers, and farmers may have an advantage in learning cob.

For workshops, classes, and volunteer learning opportunities, check:

- Cob Cottage Company, www.cobcottage.com
- regional or topic forums at www.permies.com
- trades sites like the Masonry Heater Association, www.mha-net.org
- individual builders' websites, such as the authors at http://www.ErnieAndErica.info

There are also good books and videos on general natural building, natural plasters, and cob. The following is a need-to-know summary to save some of our rural readers a trip into town.

Basic Qualities of Materials

Think of earthen masonry like a concrete or asphalt bound together with clay instead of cement or tar. Most of the volume is sand, gravel, or crushed rock, with just enough clay to hold it together.

Modern masons typically buy bagged mortar mixes or components and mix them with water. Traditional earthen masonry mostly used the range of local materials that could be dug near the building site, with minimal amendments and processing. Soils and subsoils vary widely, so local earthen building practices or "recipes" may be very different from region to region. All the historic variations can be pretty inspiring — maybe we'll just refer you to David Sheen's movie and photo gallery, *First Earth: Uncompromising Ecological Architecture*, http://www.davidsheen.com/firstearth.

The method known as Oregon Cob uses field-testing, amendments, and consistent batch testing to get reliable results in any location. While not all areas have "ready-mix" soils, most locations have the constituent materials: deposits of sand, gravel, and clay or clay-rich soils.

Quantitative Proportions

The proportions of materials in historic earthen masonry mixes vary widely. They might be:

- 6–20% true clay
- 20–55% fine and coarse sand
- 25–50% rock and gravel
- from 1.5% to 60% straw, dung, or other fibers
- other soil minerals such as chalks, silts, mica, organics, and larger rocks

Sources: World Heritage Earthen Architecture Initiative, Devon Historical Society, and other sources excerpted in *Earthen Building and the Cob Revival: A Reader*, collected by Cob Cottage Company in Coquille, OR.

When choosing dirt and deciding what to mix with it, what do you look for?

- Watch for rock-hard dirt paths as a possible indicator of good cob soils.
- Good garden soil is bad building dirt, and vice versa.
- Dig down below the topsoil (if you have any). Mineral soil is usually lighter colored, whereas good topsoil is brown or black. (Some local mineral soils may also be red, yellow, or even blue.)
- Consider digging in an area designated to become a pond or swale.
- If there is no appropriate excavation site on the land, look for local sources for rock-crusher fines, rough sand, and sticky clay fill dirt.

Identifying Clay

Clay often settles in low spots, ponds, and puddles. As it dries, the surface cracks and may curl up. Wet clay will be sticky and slick; if squeezed, it squirts through your fingers. Roll yourself a "worm" and see if you can curl it around your finger without cracking. If you can squeeze water out of it, it is not clay, but silt. Silt feels velvety in the hand, gritty if you care to sample it with your teeth. Silt is too soft and crumbly for building, whereas clay is useful, and so are sand and gravel.

Most natural clay deposits also contain some silt, both mineral and organic. Highly expansive pure clays can be mixed down with silt, mica, or extra aggregate, but non-expansive clays are easier to work with.

To buy some clay for high-grade recipes and testing, look for:

- fireclay (bagged, powdered clays suitable for high-temperature uses, like kaolins)
- pottery clay (a range of clays suitable for kiln-firing with minimal shrinkage; pottery studio "recycle bin" leftovers are excellent for most earthen building purposes)
- "mortar clay" (cheaper powdered clays with relatively little expansion and shrinkage)
- local clay-rich soils (we tend to soak and sift any wild-harvested or recycled clay before building, to soften any partially hydrated lumps).

Problem Clays

We avoid highly expansive bentonite clays, which can absorb up to 20 times their dry weight in water; they shrink back as they dry, with extreme tendency to crack.

"Fatty" or brick clays may have greater shrinkage but also greater sticking power. (In brick-making terms, a "fatty" clay would need sand added to make good bricks; ordinary "brick earth" could be used without amendment.) If you are stuck with a pond-sealer or fatty clay, remember a little goes a long way. We once got a load of lovely red brick clay and ended up doing a lot of test bricks before we found our mix: we went from our normal 3:1 sand:clay recipe up to a whopping 10 parts sand for 1 part brick clay, to get the right texture and avoid cracking in the dried project.

The cement and lime industries use powdered baked clay, called metakaolin, as a *pozzolan* (a special ingredient used to change the properties of certain cements and plasters). A few places call this metakaolin "fireclay," the same name that other places use to identify raw clay suitable for fireplace, firebrick, and furnace purposes. When buying bagged clay, if possible, test a sample by wetting it and smearing it with your fingers. Raw clay will easily make a sticky paste that covers your fingerprints. Metakaolin will act more like silt — gritty or velvety, and not sticky (water rivulets will reveal your fingerprints). Because it has already been fired, metakaolin is no longer sticky, and cannot be used as a binder in cob, earthen plasters, or fireclay mortars. It can be used as an extender in paints, plasters, or mortars, provided that you also get some raw clay to act as the binder.

Aggregates

As with concrete masonry, sharp sand and mixed-size aggregates work best. Rock-crusher fines, quarry run or "pit run," or any freshly broken sandy material such as mountain sand are excellent, and they can be sifted or screened to the desired fineness. Occasionally we get a load of sand or quarry fines that already contains some clay, sometimes in the exact proportions to make it "ready-mix" for bulk cob projects.

Bagged, graded sand can be good but is often graded for size and color, not sharpness. Beach sand and "pea gravel" are too round for most building projects; they tend to collapse like a pile of ball bearings. At lower elevations, water-worn sand may be sold as "masonry sand" in the absence of better options. If you can check a sample before the dump truck delivers, look for coarse materials that feels like sandpaper or road grade, not cake flour or seed beads.

Fiber

Where used, fibers serve as both aggregate and tensile reinforcement, and hollow straws also increase insulation value. Typical structural cob might be half or more straw by volume, or just a flake or two per batch, depending on the other materials and their purposes. (A flake is the unit that straw bales are made of, usually 2" to 4" thick, and a compressed square about 18" on a side. There's usually 12 to 20 flakes in a 2-string bale.) Besides straw, other fibers that may be used include hair, animal dung, hulls and stems of local crops, pulped wood, and synthetics like nylon, glass, or ceramic fiber.

When you look at builders' recipes, you will see proportions such as: "1 part clay dirt and 2 or 3 parts sand." This reflects the fact that most of our "clay" soils contain more than just clay. In that "1 part clay dirt" we might find 50% true clay, 40% silt, and some gravel and sand. So if you sifted the finished mix from this 2:1 recipe, you'd find about 15% clay, 10% silt, and 70% sand and gravel — right within our historic ranges for excellent cob. We have had deliveries of "clay" soils that contained pockets of nearly pure silt, and "sand" that contained so much clay it needed more sand added to make good cob. So batch testing to a consistent texture is more important than mixing with a consistent recipe.

We estimate that finished cob, like adobe, averages about 95 lbs per cubic foot. Our "thermal cob" made with particular care, without straw, or with very dense types of rock and gravel, can weigh more (maybe up to about 135 lbs per cubic foot dry weight). We base these numbers on our own experience of mixing by volume, plus values given at www.engineeringtoolbox.com.

Light-clay straw and straw-rich cob can be far lighter (as low as 40–60 lbs per cubic foot for infill materials, and we've recently heard of far lower values in the range of 15-20 lbs per cubic foot for light-clay insulating straw bricks). These materials are used more for insulation than for heat storage. Straw is less conductive than all the other materials in cob. If dense thermal mass and tensile strength are both wanted, another fiber such as horse dung, or shredded jute, rope, or burlap could be used instead. For our low benches, we find that a fiber-free thermal core works just fine, with a fiber-enhanced outer plaster for harder-wearing contact surfaces.

Cob can handle occasional surface wettings: the clay swells to slow water penetration. The core moisture levels must remain below about 13% to 15%, like dry firewood, to ensure structural integrity. (Source: Mike Wye & Co, www.mike-wye.co.uk) It is hard to measure core moisture accurately, but persistent damp patches, mold, rotting straw, unpleasant odors, or in extreme cases water erosion, all can indicate a serious problem. However, unlike the 2- to 14-story cob walls that most builders are working with, our 18″ bench probably won't collapse due to a little moisture — unless someone steps directly on a pipe while the whole thing is soggy-wet. For outdoor projects, a roof and damp-resistant footings are strongly recommended; see the "Greenhouse" section in Chapter 3 for other damp-resistant detailing options.

We make several different types of earthen materials for a typical rocket mass heater. Some good pictures of the cob-mixing process can be found online. For simple line drawings, see Becky Bee's *Cob Builder's Handbook,* http://weblife.org/cob/cob_049.html.

Good mixing is essential — unmixed clay will crack. If possible, start hydrating the clay weeks ahead of time. Fully hydrated clay is easier to use in all stages of cobbing, and letting it stand in water is arguably the least possible work to achieve full hydration. (For those, like us, who don't always plan that far ahead, it's also possible to speedily hydrate the clay by grating it into water, and/or using power-mixers. Mortars and cob will improve if left covered overnight, like cookie or pie dough improves by a few hours in the fridge.) For layers with fiber, the fiber is added last, after the clay and sand are thoroughly mixed.

Clay Slip

Clay slip is simply clay and water, mixed to a liquid suspension. It is used as thin-set mortar, surface primer, and ingredient in other mixtures. Always good to have around. The following is a rough recipe.

> ### Clay Slip Recipe
>
> Fill a 5-gallon bucket about ½ full of water, then add a similar amount of powdered or screened clay. Break up any big lumps; soak the clay overnight if possible; then mix with hands or a paddle mixer. (If we have

any concerns about dangerous debris from previous activities on site (broken glass, e.g.), we use heavy gloves or tools and pass all clay through a ¼" mesh before handling.)

For dry clay: ⅓ to ½ bag of powdered fireclay (about 20 lbs dry weight) per 5-gallon bucket of water. Caution: Avoid breathing clay dust — protect mouth and nose while mixing.

For wet clay: Crush lumps with tools or rocks, or grate damp lumps, like cheese, through a ¼" screen into the bucket until the pile just pokes its tip out of the water. (If the dry clay soaks up your water, add more water.) Allow to soak as long as convenient before mixing.

Good creamy slip feels slightly sticky to the touch, coats a spoon or hides your fingerprints, and retains dimples on its surface when splashed around.

CLAY-STABILIZED PERLITE RECIPE

1 bag of horticultural or building perlite (4 cubic feet) + 2–3 gallons of clay slip.

Moisten the perlite with a sprayer hose or watering can to keep the dust down. (If the bag is soggy, drain off excess water before adding clay.)

Pour the perlite onto a tarp (8' x 10' size), without breathing any dust.

Drizzle clay slip over the perlite: a scant half-bucket of clay slip (2 or 3 gallons) per 4-cu-ft bag of perlite.

Roll the mixture back and forth in the tarp. You can stomp if you like, but I usually mix it with gloved hands to completely coat the perlite with clay.

To do a batch test, make a ball of clay-perlite mix. It should press together. Then pinch the ball between one finger and thumb. You want the mix still dry enough to burst apart when pinched. If too dry, add a little more water or clay slip. If too wet, allow it to dry in the sun, or add some dry perlite if available; then remix and retest.

Clay-stabilized Perlite

We use clay slip to bind the perlite together. We make it very dry, with minimal clay, to preserve the insulating air spaces inside the perlite.

CAUTION—RESPIRATORY HAZARD: Moisten the perlite before releasing it from the bag, and/or use respiratory protection to avoid breathing the extremely fine, sharp particles of perlite dust.

Clay-sand Mortar

Mortars are used to bed unit masonry, such as bricks, and to create sealed joins between masonry and other materials. Earthen mortars are particularly heat tolerant.

Mortar ingredients should be screened before mixing to remove any large particles. Use ⅛" mesh or window screen, or purchase pre-screened materials by the bag or yard.

> **BASIC CLAY-SAND MORTAR RECIPE**
>
> Successful mixes range from straight sandy-clay soil (just mashed up a bit), to equal parts sand and light-clay soil (1:1), up to 10 sand:1 clay for fat brick clays.
>
> An average mix has 3–4 parts sand to 1 part hydrated pottery clay, fireclay, or sticky clay slip
> OR
> 2–3 parts sand to 1 part hydrated local clay-rich soils (very sticky, but not quite pure clay).
>
> Water to appropriate texture — a putty-like consistency somewhere between cake batter and cookie dough; test your first batches by practicing with some spare bricks.
>
> Mix thoroughly.
>
> Earthen mortar often improves if allowed to stand overnight, allowing the clay to fully hydrate. If you notice any settling (sand on the bottom, clay or water on top), remix briefly before use.
>
> To do a batch test, lay some bricks with it. Mortar should be stiff enough to hold a shallow ridge a few inches tall (soft to almost-stiff peaks), but loose enough to splooge out when tapped or compressed. Any finished mixture should be uniform in appearance: no streaks or lumps of unmixed material.

Traditional mortars are not "brick glue" — they do not hold bricks together. They hold bricks *apart*, like filling putty or tiny wedges. The mortar's job is to help you seal and level the bricks.

To see how mortared courses work, lay a few damp bricks, compare them with a level, and then tap the high ones down into the mortar. The mortar should be soft enough to allow you to work on four to eight bricks at a time. Dip each brick in water or weak clay slip before setting to ensure a good bond with the clay-based mortar.

Our old 1977 edition of the *Audel Masons and Builders Library* gives this description of mortar consistency: "The amount of water in mortar is more a matter of experience than any rule. Since some water is lost due to absorption by the brick, some allowance is made for this. More important is the effect on the ease with which mortar is applied to brick when a wall or other structure is laid up. The water content must not be so great that the mortar slides off the trowel when picked up, nor oozes off the end of a brick when it is laid up." (pp. 29–30, *Bricklaying — Plastering — Rock Masonry — Clay Tile* by Louis M. Dezettel, published by Howard W. Sams & Co, Inc., Indianapolis, Indiana, 1972. Mr. Dezettel is describing Portland cement mortars, but the same consistency and workability applies to all kinds of mortars, including lime-sand or clay-sand.)

Remember you want an airtight seal on these projects as much as you want a level course. You can "butter" each brick to make a long ridge by scraping a trowel across each edge at an angle, parallel to the joint. When you set the brick, tap it into place to compress the ridge, to make a guaranteed seal in middle of the joint where the ridge was. You may need to add more mortar to one side or the other if the ridge squished unevenly, but you will not have hollow bubbles in the center just waiting to let gas leak out.

Leftover clay mortar in the bucket may be re-wetted and re-used, unlike most other types of mortar. But don't try to reclaim the scraps that drop off the project — they may include lumps of gravel, debris, dust, and other unwanted materials. Instead, sweep them and dump them into the rough cob mix or gravel pile.

Cob: Thermal and Structural

Cob is very similar to earthen mortar, but is usually made with rougher materials, more aggregate, and a stiffer consistency (less water). Cob is by definition building with lumps of pliable earth, without formwork; in order to build this way, the material needs to be stiff enough to support its own weight (6″ to 18″ of wet material can be stacked in the same day, without formwork, if the cob is stiff enough).

We use a basic earthen building mixture without straw for the core of our benches. We call it "thermal cob" to distinguish it from "structural cob," which contains straw or long fibers.

Straw is pretty much non-flammable when encased in clay-sand material; it may char slightly but you'll never know it except for a faint smell. If you happen to have leftover straw-cob from another project, it can be used in the bench core. However, because straw's hollow core gives it high insulation values, too much straw can reduce heat transfer in the thermal mass. Short benches don't need that much tensile strength anyway, so the straw is more work to no purpose.

Straw or fiber is useful in outer layers that see some wear and strain, and possibly as insulation in low-heat areas (if it's cheaper than perlite).

Cob Test Bricks

Before mixing large batches of cob, it pays to check your materials and proportions. Most suppliers will give you a sample of their masonry sand or other materials — all you need is about one 5-gallon bucket or sandbag — to combine with your local mineral subsoil and determine the final quantities.

Use pint or quart yogurt containers instead of 5-gallon buckets for your test batches. You can mix the test batches by hand or foot. Use as much local mineral soil as possible. (Mineral soil is the lighter-colored material beneath the dark topsoil, without many roots or wrigglers.)

Make test batches with the following proportions:

Sand	:	Clay
0	:	1
1	:	1
2	:	1
4	:	1
6	:	1
10	:	1
1	:	0

(We added the "zero" mixes after we worked on several sites where the "sand" contained enough clay to be used without further amendment. If you can find such ready-mix — sometimes available in the ground, sometimes from rock-crusher fines or stone quarries — it's ideal for making cob.)

If your soil is clay-rich, use it in place of the clay (smaller quantity), and add purchased sand in the larger amounts as above. If local soil is sandy, use it in place of the sand (larger quantity), and add a small amount of high-grade ceramic clay or fireclay.

Mix all test batches with water to roughly the same consistency (like cookie dough or stiff paste). Shape each batch into a large brick, about 8″ on a side by 2″ thick. Use the same forms for each brick if you want to compare shrinkage. Mark the bricks with chalk or finger-dots. Take a picture and/or note the batch recipes in your project notes, so you know which brick represents which mixture.

Dry the test bricks in the sun, or even in an oven. Yes, this might make them crack. But if they're going to crack under heat or pressure, you want to know NOW, not once you have mixed three tons of the material and layered it into your living room.

Cracking may indicate too much clay. Crumbling or "dusting off" may indicate too much sand or silt in the mix. Both cracking or crumbling may indicate not enough mixing. It is structurally important to break up the clay lumps and completely coat the sand particles with clay.

Shrinkage from original dimensions can also indicate too much clay, not enough sand or aggregates, or possibly excess water. A small amount of shrinkage is normal (less than 1/16 inch per foot). Less-expansive clays are easier to work with since they don't shrink as much, even when used to excess.

If local soil is very silty or rocky, without a significant amount of sand or clay, you may still be able to use it to bulk up batches of locally purchased materials. We have done mixes with 2–3 parts trucked-in sand, 1 part local soil (silt/clay), and ¼ part purchased clay, in order to make the best of a silty site.

Even with industrial vehicles and fuel, it is not usually worth trucking tons of materials more than a few miles to make cob. If the local soil is mostly silt or rock, consider fieldstone or dressed stone masonry instead of cob. A few bags of clay will make enough clay-sand mortar to bed a mostly stone bench and create a good seal around the pipes.

Mixing Cob

Can "cob stomping" be mechanized? Yes, but it's not always worth it. See Chapter 4. Power tools add danger and expense to the

> **COB BASE RECIPE**
>
> 1 part local clay soil
> (optional: up to ¼ part additional fireclay slip, or pottery clay slip, if needed)
> 2–4 parts aggregate (or as determined by test batches above)
>
> Generally this will be 5-gallon buckets of sand and soil, with quarts or up to a gallon of refined clay slip.
>
> Roll the mixture back and forth on an 8′ x 10′ tarp until the dry materials are roughly combined. Then tramp or tamp the material, using sand and gravel to break up any large clay lumps. Roll again, tramp again, a couple more times. Add enough water to make a cookie-dough texture — soft enough to squash but stiff enough to pile. (I generally run the hose until I see water just starting to pool in the footprints, then mix that in, then add more if needed.)
>
> Continue mixing until it passes the batch tests (below). Then you can add more straw or additional sand.
>
> Stomp and roll the cob. Remember: *you only need to step on each place once,* like kneading bread. Roll it over to expose fresh territory, and stomp again. Music improves morale. When almost ready, the cob will begin to hold together when you roll the tarp, like dough. This is a good sign. We sometimes call it "the Seal of Approval."

work site, and with inexperienced operators, mixing by foot can actually be faster.

If you plan ahead, you may be able to work "accidental cob mixing" into your ordinary farm activities, for example adding sand to a muddy clay-bog in a problem driveway before the rainy season, then carting the whole lot away to your cob project site and replacing it with the drainage gravel and road-grade rock that you actually need in that spot.

Batch Testing Cob

We check each batch for consistency with the following simple, wet tests.

1. *Changeability test:* Work it in your hand to form a ball. Does it get stiffer, stickier, or better as you work it? If it changes, the batch needs more mixing. There should be no visible streaks or lumps.
2. *Drop test:* Drop the ball onto the tarp from 4′ (about shoulder height). It should remain a ball, or compress slightly into a domed lump. No crumbling sand piles.
3. *Squeeze test:* Take a handful in a tight fist, feel and listen for the gritty sound of jamming sand grains. Note any excess clay that tries to squeeze out. (I like to catch the squeeze-out, and then squeeze it in my other hand. If the squeeze-out is gooshier than the main mix, the batch can take more sand or straw.)
4. *Smear test:* Is it sticky enough? Spread your fingers like laths for plaster. Smear some of the mix onto your spread fingers, and turn your hand over. Can it hold its own weight upside down? Open and close your hand slightly (an inch or two). With bare hands and a very good batch, you should be able to flex your hand 5 to 10 times before the material starts to fall off. Thermal cob may only stick through 3–4 flexes if it already has its full sand content.

Once the batch passes these tests, you can add the last parts of aggregate.

- For Thermal Cob: Add 1 more part of sand or gravel to make a stiff, rough cob for infill areas (bench core, fireplace or brick oven backing).
- For Structural Cob: Add straw — a flake or two per 4-bucket batch, or about 2 cubic feet of loose straw — and thoroughly mix it into the earthen materials. The more straw, the better, for most purposes. When the material starts to fall apart, stop adding straw — or dip the straw in clay slip if you need to keep adding more.
- For Earthen Plaster/Rough "Brown-Coat" or Scratch Coat: Add chopped straw, chopped hair, or horse dung to make plaster. Rough plaster is good for filling gaps and cracks in the original material. For fine plasters, add screened aggregates and fiber first, and make the mix a little wetter than cob for better smearing action.

Unlike the mineral components of cob, natural fibers will decay over time if left damp. Batches that sit damp for more than a week should have fresh fiber worked into them before building. Batches that have so much rotted fiber they resemble compost should be sent to the garden, not used in a building project.

Earthen Plasters

There are a wide range of combinations that work, from sand-clay with a whisper of hair fiber, to dung plasters or ditema/litema that may be 60% to 90% fiber with just a hint of clay.

Sift all materials well, to eliminate particles bigger than one third the thickness of the intended plaster. (For a ½-inch plaster layer, use ⅛-inch hardware cloth or window screen, or bagged and graded "sugar sand.")

- 1 part mortar mix (sifted sand and fine clay)
- 1 part soaked and separated fiber
- If needed, extra clay slip, binders/hardeners, tints, and amendments to suit.

Use chopped and separated fiber such as:

- chopped straw/short straw animal bedding, sifted through ½" mesh.
- pet hair or chopped hair trimmings
- chopped fiber from old natural-fiber rope, or "oakum" (not tarred)
- cellulose fiber pellets, as sold for concrete mixes
- horse dung, fresh or dried (or dung from cow, donkey, goat, sheep, elephant, or any other critter that eats mainly dried hay or sticks)

Squick Factor: It's fine to use other materials if you don't like the idea of touching poop. But it works so well, and is available so cheaply, and gives such great results that it's worth a try. (There is no scent after it dries, and not much while working with it in the wet clay plaster.)

Some safety tips if this is your first foray into handling dung:

- Use rubber gloves, or even garden gloves to prevent abrasions. Your skin is a great germ barrier, as long as it doesn't get scratched up.
- Never use scat from animals that eat meat or slops, or that are sick.
- Avoid aged or rotted manures; if it looks like topsoil, or has things living in it, send it to the garden.
- Be sure to wash thoroughly with soap and water before eating or handling food, of course.

Fine plasters often have additional ingredients to beautify or harden the surface:

- mica, rice hulls, or other "glitter" and extenders
- chalk, whiting, or pigments such as concrete pigment, paint tints, natural colored clays
- white or colored sand for more color options
- hardeners (but remember, more is not better):
  ○ wheat paste (1 cup per 5-gallon bucket, or about a pint per 3-bucket batch),
  ○ yogurt or milk products (about 1 pint per 3-bucket batch)
  ○ linseed oil: a spoonful per bucket (too much can interfere with the clay's bonding)
  ○ up to 3% lime putty, or use lime-water instead of plain water.
- mold retardants/binder softeners: borax or lime.

Hardeners: More Is Not Better

Overusing hardeners can weaken the mix.

Some builders think adding cement will improve the situation. "Cement-stabilized" earthen walls are no longer a clay-based material, they are very poor concrete. (See Graeme North white papers, at www.ecodesign.co.nz for more information.) Clay swells when wet, sealing the surface against moisture penetration. If there is enough lime to interfere with the clay, but not enough to replace the clay as a surface protecting layer, the combination can let water penetrate further yet dry out more slowly.

Dairy casein, wheat paste, or cactus gel will swell with moisture like clay does, so these gel-type hardeners are clay-compatible and can be used in larger amounts (1 or 2 cups per bucket). Their only limit is the possibility of a too-rich material that peels instead of bonding (like any rich paint over a soft surface), and their nutrition value may attract pests or mold.

More is not always better. Use proven recipes, or test well.

For plasters, experiment with finish techniques on a dried section of the surface. Watch for cracking, dusting off, and the gloss and colors you want for the finished product.

Applying Cob

For the building process, work thick cob in courses, like rough bricks or inverted Legos. Use downward, poking, and smearing motions. (Don't pat it into pudding, it weakens the matrix.) If working new material onto dry surfaces, use water or dilute clay slip to moisten the dry surface and improve the bond.

Think in terms of flat courses for walls or benches, angled or "keystone" courses for domes and arches, and corbeled or cantilevered courses for overhangs and ledges. But mostly, if in doubt, keep the courses flat, just as if the cobs were bricks or adobes. At any point, you should be able to walk your fingers along the finished layers of project as if they were tiny little inspectors, and they should not have to perform feats of mountaineering or balance.

Work with a downward pressure, supporting the shape from the side. If the cob starts to slump, splooge, or fall outward, never push it back — this creates a wobbly joint that is prone to slump again.

Slumping usually indicates that you are as high as you can go until the cob dries. Stop working on that area and move along. Come back in a few hours, when the material has stiffened somewhat, and trim off any excess with a trowel (using a downward or sidelong motion). Allow the material to firm up substantially before adding more weight on that section. The material you've removed can be used elsewhere, or mixed into a new batch of cob. In fact, all the grit and earth that land on the floor tarps can be mixed back into a new batch, as long as there's rough cob going into the project somewhere.

Incorporate as much rubble, gravel, and broken brick as you can get your hands on. It's all thermal mass; just make sure it's thoroughly coated in clay slip so it sticks to the cob. Keep rubble back a few inches from finished surfaces for easier trimming.

When the cob is "leather hard" (firm, but cool and dark), step back and evaluate the line of your project. Use an old machete, flat shovel, or a "cob rasp" (expanded-metal lathe over a 2x4 wood block) to trim away any bulges or proud spots, and bring the face plumb and level.

You may see some cracking or shrinking as the whole mass dries. Allow the mass to dry completely, and check the cracks for structural soundness.

If the material is fully adhered and sealed, but just separated at the surface, you can just let it dry fully and then plaster over it once it has stabilized. If the cracked area is large, crumbling, or unstable, you will want to remove any bad material and fill the gaps with a good, stiff, sandy mixture.

See Chapter 5 for more detail on repairing cracks and finishing options.

Finished Surfaces

There are several ways to finish earthen masonry.

Burnishing:

The simplest method, if you love the color and texture of your cob, is to use floats and trowels to smooth the surface. Remove any straw-joints, protruding rocks, etc., and fill in any low spots. As the cob stiffens, wet-float over each area to further smooth the straw and aggregates, and burnish clay off the surface to expose the other materials' color and visual texture. When completely dry, finish by burnishing with a curing oil (linseed or walnut, e.g.).

Two- or three-coat plaster:

After the cob is dry, make a rough plaster to fill in any divots or hollow areas of the wall. Moisten the cob with clay slip before plastering.

Apply the plaster with a trowel or float, in big sweeping motions of the full arm. Bring the surface as close to plumb as possible with the first coat. Work the plaster firmly into any dents, divots, or thumb holes to adhere it.

It's sometimes easier to use two or three thin coats than one thick one — very uneven plaster thicknesses may crack.

Scratch up the previous coating of plaster before moistening and adding the next coat.

You may want to practice tinting in your second coat, in case you feel like stopping there. If you do need a third coat, it can be thinner, and will wear better for being laid on over a matching-colored base.

The finish plaster can also be lime, gypsum, or any breathable mineral plaster. Avoid anything with latex, acrylic, or polymer ingredients, and avoid Portland cement stuccos and synthetic grouts.

Wood trim:

It's best to plan ahead and install "dead men" for wood framing before the cob is cured. Dead men are pieces of wood with prickly branches or spikes on the cob side, and an exposed face for framing on the surface side. The spikes are buried in wet cob, worked in well. The framing face is left proud (sticking out above the surface) for easy framing and finish plasters. Wood and other materials can be attached to the "dead man" after the cob is dry. (See Chapter 4, "Wood Trim.")

Make sure to leave good clearances between heat sources and combustibles. Cob is a non-flammable material, but buried wood can still be damaged by heat.

Tile work:

Any tiles for earthen masonry should be set in a mineral grout, such as earthen mortar, lime putty, or a lime-sand plaster. Lay out the design ahead of time, as for any tile project, then wet or clay-slip the backs of each tile when pressing them into wet plaster. Allow them to set, and then apply a top-dressing of grout and wipe clean to create a smooth, level, finished surface.

Earthen masonry benefits from remaining "breathable" — and so will your household, since this breathable feature helps stabilize indoor humidity at very comfortable levels. We often recommend using tile for detail, or for a vertical or horizontal surface, but not both. Breathable earth or lime plasters on the remaining surfaces can be an attractive and healthy finish.

If you do want an entirely tiled surface, using mostly small tiles with reasonable amounts of breathable grout in between can allow enough transpiration to prevent moisture buildup or blistering behind the tiles.

Stone masonry or brick facades:

Masonry facings can be built into the structure course by course, or built up separately and filled with earth or mortar. If the unit masonry facing work is structural (supports a seating slab to take weight off the pipes), then you can consider the earthen mortar as "fill" rather than structural cob, and use your preferred masonry materials and methods.

Field stone facings can be worked in place during cob construction. This gives maximum integration between the visible stone facing and the interior rubble and cob. Long, skinny tapers that extend well back into the earthen masonry help tie the facing and fill together.

Flat stonework can be added afterwards, like tile, on the seats and other flat horizontal surfaces. Cob doesn't support "glue-on" rockwork on vertical surfaces, so decide in advance if you want a rock front, and build it that way.

Brick facades can be built first and used as formwork for the earthen materials. You can expect slow settling and slow drying of cob trapped between brick walls.

If you want to integrate modern, Portland-cement-type materials in the casing, please read the references by Graeme North and Mike Wye, below, for detailing considerations. You may need to give extra attention to thermal expansion jointing as well as moisture transport.

Ensure that any enclosed earthen masonry or brickwork is completely dry and protected from further moisture (including ground damp, condensation, weather, wicking from the cement materials, internal exhaust condensation, and spills).

Cement based materials tend to trap moisture within adjacent natural materials, potentially leading to degradation, cracking, or mold issues. Expansion jointing will be critical around the barrel and firebox if using rigid masonry (Chapter 4 notes several points during installation when expansion jointing should be considered, and Chapter 5 gives tips for after-the-fact repairs).

Natural paints:

After burnishing or plastering, you may wish to freshen up or tweak the color.

Paints for natural materials should be compatible; they must be breathable — able to penetrate into porous material. If the paint is too strong, it will bond to itself but not the surface, and eventually peel. If it is too watery, it will soak into the surface and require more coats to cover.

Common natural paints include:

- Clay paint: 1 part clay slip, 1 part pigment or extender (e.g., talc, chalk, or mica)
- Aliz: 1 part clay, 1 part mica, up to ¼ part wheat paste (or use a micaceous clay)
- Milk paint: milk base: 1 part milk curds or plain yogurt, 1 part saturated borax-water; heat and stir. Add 2 parts pigment plus extender to make paint. Save pigment

separate from binder. Water as needed to prime porous surfaces. Makes a very permanent bond on woodwork.
- Lime paint/whitewash: 1 part milk base, 2 parts lime putty, pigments and water to suit. Often used as a very watery wash; it goes on translucent but brightens to white as it dries. Not very suitable for damp areas, but repeated coatings can create a soft, lustrous white wall.
- Egg tempera: Contents of 1 egg yolk, plus an equal amount of water is enough binder for a few tablespoons of pigment/extender. Add water for washes or glaze effects.

Painting on old plaster or earthen material, one must take care that the paint is thin enough to penetrate and bond. When working paint onto dried surfaces, do not moisten the surface — it can cause surface material to loosen and roll up into the paint. Instead, paint directly on the dry surface. If needed, water down the paint before applying to make a soaking-in primer coat, and allow to dry before going back over the same area. Brushes and rags often work better than rollers. Once the paint is mostly dry, you can also go back and burnish with a rag or paper to bring out different sheens and textures.

With natural paints, multiple coats are often needed to produce matte, even effects. There's a balance between pigment loads and penetration. If you can get good, finely ground pigments from a paint supplier or art supply, you will get better paint coverage with the same appropriate binder and water ratios.

Instead of trying to replicate the white anonymity of drywall, consider deliberately working with a consistent, artful stroke texture such as small circles or cross-hatching. Many people pay big bucks for faux painting effects that mimic dappled, old-world materials. You are working with an ancient material; it's OK to let the "Mediterranean" or "Old World" textures show through. If you don't like the effect, you can always paint over it.

Detailing for Earthen Structures

Any building needs a good hat and a good pair of boots (roof overhangs, and well-drained foundations) to ensure the long life of natural materials. With breathable weather protection, earthen structures can last many centuries.

We never use vapor-blocking "Band-Aids" like plastic sheathing or concrete stucco with earthen masonry. There are too many sad stories about an earthen wall that stood for centuries, then got "protected" with concrete stucco, and collapsed within a few years from the moisture trapped inside the stucco. It is difficult to tailor a vapor barrier so that it can handle both interior moisture migrating outward, and exterior moisture migrating inward. Breathability is key for cob.

References and Resources

Cob and Earthen Building Methods

Cob Cottage Company in Oregon: www.cobcottage.com, see also *The Hand-Sculpted House,* by Ianto Evans, Linda Smiley, and Michael Smith, published by Chelsea Green.

The Cob Builder's Handbook, by Becky Bee http://weblife.org/cob/.

Clay Culture: Plasters, Paints, and Preservation, by Carol Crews, Gourmet Adobe Press, Rancho de Taos, New Mexico, 2010, distributed by Chelsea Green. A good general introduction to earthen building, particularly the US Southwest methods, and great finish recipes. See www.carolcrews.com.

General Information/Historic and Heritage References

Graeme North in NZ (www.ecodesign.co.nz) — wet-climate cob code and detailing recommendations for outdoor protection.

Mike Wye in the UK (www.mikewye.co.uk) — numerous informative articles on historical detailing and appropriate renovation or restoration methods.

World Heritage Earthen Architecture Program http://whc.unesco.org/en/earthen-architecture/ — well-researched detail on ancient, traditional, and modern earthen building in various places.

First Earth — a documentary film about ecological architecture. See http://www.davidsheen.com/firstearth/film.htm.

The Book of Masonry Stoves: Rediscovering an Old Way of Warming, David Lyle, 1984, Brick House Publishing Co, Inc. Chelsea Green, White River Junction, Vermont, 1997. A wonderful introduction to the history of masonry heating, with detailed plans of historic examples. Also distributed by the Masonry Heaters Association, www.mha-net.org.

While you are browsing the Internet, try the following search terms for inspiration: Shibam in Yemen, Taos Pueblo and Mesa Verde, the Great Wall of China, Dracula's Castle Stove, Pyromasse, kachelofen/kakkelofen, contraflow, k'ang, ondol, tawakhaneh, Swedish candle or Swedish torch.

Workshops and Local Builders

Since the recent closing of the Natural Building Network website, there is no single site with the sole purpose of professional listings for natural builders. Workshops, skilled builders, and local project coordinators can sometimes be found through the following online sources:

- Permies (permaculture discussion forums): www.permies.com/forums.
- The Cob Cottage Company calendar, www.cobcottage.com.
- Authors' calendar: www.ErnieAndErica.info/upcoming_workshops.
- Leslie Jackson often lists workshops: www.rocketstoves.com/workshops/.

The Masonry Heater Association includes some members who are interested in rocket mass heaters, and many talented and certified masons with their own specialties: www.mha-net.org.

Appendix 2:
Rocket Mass Heater Building Code (Portland, Oregon)

ORIGINAL TEXT AND COMMENTARY is available at: www.portlandoregon.gov/bds/article/214146, article 09-002 "Rocket Mass Heaters: Final Recommendation."

[Author comment: This code was originally drafted in 2009, through a collaboration between Ernie and Erica Wisner, Joshua Klyber, Bernhard Masterson, and other natural builders. It was approved for use with a one-week permitting process by the Alternative Technologies Advisory Committee, of which Mr. Klyber was a member, in 2013.

A few elements in these guidelines do not agree with our current recommendations, and we've inserted author comments pointing them out where they appear below, in case readers are considering using these codes as a model for a local code appeal or amendment. These author comments are not part of the legally-approved code guidelines.

Only the building guidelines have been reproduced here; for the full supporting documentation, including comments on building code conflicts, typical materials, history of use and testing, and other factors, please see http://www.portlandoregon.gov/bds/48661?a=437516]

1. Scope:

1.1 This guide covers the design and construction of Rocket Mass Heaters, a subset of solid fuel burning masonry heaters. It provides dimensions for site constructed rocket mass heaters and clearances that have been derived by experience and found to be consistent with safe installation of those rocket mass heaters.

1.2 Values given are in English measurements, and are regarded to be standard. All dimensions are nominal unless specifically stated otherwise. All clearances listed in this guide are actual dimensions.

2. Definitions:

2.1 Combustion unit: The area where fuel is consumed and clean exhaust produced; comprising the fuel feed, burn tunnel, heat riser, barrel, and the manifold.

2.2 Combustion unit base: Area composed of the fuel feed, ash pit, burn tunnel, insulation and casing.

2.3 Fuel/air feed: Area where fire is lit and fuel is added. This is the sole air intake.
2.4 Ash pit: Optional depression located at bottom of fuel air feed and/or manifold.
2.5 Burn tunnel: Horizontal area where initial combustion occurs.
2.6 Heat riser: Internal chimney, insulated for high-temperature combustion and draft.
2.7 Barrel: Metal or masonry envelope around the heat riser that radiates heat.
2.8 Manifold: The connection between the combustion unit and the heat exchange ducting.
2.9 Heat-exchanger: The volume of mass that absorbs heat from the heat exchange ducting and re-radiates it over an extended period of time. Comprised of the heat exchange ducting, thermal core and casing.
2.10 Heat exchange ducting: The flues that carry hot exhaust gas through the thermal mass.
2.11 Thermal core: Area directly around heat exchange ducting.
2.12 Flue exhaust: The portion of the ducting after it leaves the thermal core.
2.13 Casing: Durable external layers to protect thermal core, provide additional thermal mass, maintain desired surface temperature, and allow decorative expression.
2.14 Cleanout: Capped opening for maintenance access.
2.15 Damper: A duct valve that regulates airflow.
2.16 Make up air: Air that is provided to replace air that is exhausted.
2.17 Combustion air: The air provided to fuel burning equipment for fuel combustion.

3. Guidance and use:

3.1 This guide can be used by code officials, architects and other interested parties to evaluate the design and construction of rocket mass heaters. It is not restricted to a specific method of construction, nor does it provide the principles to be followed for the safe construction of rocket mass heaters.
3.2 This guide is not intended as a complete set of directions for construction of rocket mass heaters.
3.3 Construction of rocket mass heaters is complex, and in order to ensure their safety and performance, construction shall be done by or under the supervision of a skilled and experienced rocket mass heater builder.

4. Requirements:

4.1 Sizing:
    4.1.1 6″ flue rocket mass heaters can be installed for any heated space 1000 sq. ft. or less.
    4.1.2 8″ flue rocket mass heaters are typical and appropriate for any installation.
    4.1.3 Cross sectional area shall remain consistent throughout the system except in the barrel and manifold, where it may be larger.
    4.1.4 Ducting may be tapered to reduce diameter by 30–40% in the final third of its length to improve gas flow. *[Author comment: This may have been a misprint, since reducing the diameter 20% reduces the flow area over 40%. Not recommended to reduce the diameter more than 10–20% for most systems.]*
4.2 Foundation: The combustion unit base and heat exchanger shall be supported by a concrete slab or equivalent. Any alternative configurations or foundations shall be engineered.
4.3 Combustion unit: Shall be constructed in either earthen masonry or refractory

materials. If refractory materials are used, an expansion joint shall be included between combustion unit and earthen masonry.

4.3.1 Mortars, when used, shall be clay-sand, fireclay or suitable refractory mortar. Mortars may be omitted if using monolithic earthen masonry as an external seal.

4.3.2 Fuel feed:

4.3.2.1 Shall be constructed of dense firebrick, clay brick or equivalent refractory material able to withstand 2200°F.

4.3.2.2 Shall have a closure that serves as an emergency shut down lid.

4.3.2.3 Shall be the sole air intake.

4.3.3 Ash pit:

4.3.3.1 If included, shall be located at bottom of fuel feed.

4.3.3.2 Shall be of a depth no greater than 4" below the burn tunnel.

4.3.4 Burn tunnel:

4.3.4.1 Shall be constructed of dense firebrick, clay brick or equivalent refractory material able to withstand 2200°F.

4.3.4.2 Shall be insulated with 2" of clay-perlite insulation or equivalent underneath, above, and on all sides except the fuel feed and heat riser openings.

4.3.5 Heat riser:

4.3.5.1 Shall have a minimum height of twice the burn tunnel length.

4.3.5.2 Shall be at least three times the height of the fuel feed.

4.3.5.3 Shall be constructed of firebrick, clay brick, metal flue or equivalent refractory material able to withstand 2200°F.

4.3.5.4 If constructed of metal flue, it shall be free from holes, wrinkles, burrs, jagged edges or other obstructions. Metal shall be high-temperature stovepipe or steel.

4.3.5.5 Shall be insulated with minimum of 2" (or the maximum amount that does not restrict airflow) of clay-perlite insulation or equivalent.

4.3.6 Barrel:

4.3.6.1 Shall be free from holes, wrinkles, burrs, jagged edges or other defects.

4.3.6.2 Any existing paint or surface coatings shall be removed. High-temperature coatings rated for woodstove application such as stove enamel, cast-iron seasoning oils may be used.

4.3.6.3 A cleanout shall be located near the base of the barrel that allows access to the manifold and initial ducting, or the barrel shall be configured for removal.

4.4 Heat exchanger:

4.4.1 When the heat exchanger mass is made of earthen masonry it shall rest on a stabilized masonry base with a minimum height of 2 inches.

4.4.2 Ash pit: If included, shall be located directly below the manifold and within easy access of the cleanout.

4.4.3 Ducting:

4.4.3.1 Shall be metal flue, ceramic flue liner, well-pointed brick, earthen block, or equivalent alternative. Metal flue shall be free from holes, wrinkles, burrs, jagged edges or other defects. Metal flue shall be at least 26 gauge.

4.4.3.2 Shall be embedded in a continuous layer of earthen masonry for both thermal contact and gas seal.

4.4.4 Cleanouts:

    4.4.4.1 Shall have a sufficient number of cleanouts such that all sections of the ducting shall be accessible.

    4.4.4.2 Cleanouts shall be the same minimum size as system: e.g. 6" diameter for a 6" flue rocket mass heater and 8" for an 8" flue rocket mass heater.

4.4.5 Thermal core shall be encased with a minimum of 2" thermal earthen masonry around the ducting. *[Author comment: See Chapter 5 for our current recommendations for safe seating temperatures and clearances — a 5" minimum thickness.]*

4.4.6 Casing:

    4.4.6.1 Total thickness to surface from ducting, for a 6" system shall be at least 4" depth around the first 10 feet of ducting, and 3 inches around the remainder.

    4.4.6.2 Total thickness to surface from ducting, for a 8" system shall be at least 6" depth around the first 10 feet of ducting, and 4 inches around the remainder.

4.5 Flue exhaust:

    4.5.1 Shall be composed of metal flue, clay flue lining or equivalent.

    4.5.2 Shall be equipped with a ferrous metal damper and shall be operable from the same room as the rocket mass heater.

    4.5.3 Shall have a flue priming port or other means to ensure proper drafting.

    4.5.4 Exhaust termination:

        4.5.4.1 Exhaust termination shall be at least 2' higher than any portion of a building within 10'.

        4.5.4.2 Exhaust termination shall be protected from precipitation and vermin.

4.6 Make-up air: *[Author comment: We consider air pressure problems to be a building issue rather than a heater issue. This code does not distinguish between dedicated exterior air, "outside air," and make-up air, nor does it address safety issues in outside air construction. See Chapter 6 or Appendix 3 for more about outside air.]*

    4.6.1 Rocket mass heaters shall be supplied with an exterior air supply unless the room is mechanically ventilated and controlled such that the indoor air pressure is neutral or positive.

    4.6.2 Exterior air supply:

        4.6.2.1 Shall be capable of supplying all combustion air from outside the dwelling or from spaces that are within the dwelling that are non-mechanically ventilated with air from outside the dwelling.

        4.6.2.2 Shall be in the same room as the rocket mass heater.

        4.6.2.3 Shall not be located more than 1' above the height of the fuel feed. *[Author comment: This is consistent with other codes, and practical for make-up air to room. For dedicated outside air, constraints are different for this type of heater. Our first successful outside air installation is described in Appendix 3 of this book.]*

        4.6.2.4 Exterior air supply inlet shall be protected from precipitation and vermin.

        4.6.2.5 Exterior air supply shall have a shut off valve to prevent cold air infiltration when not in use.

4.7 Clearances:

    4.7.1 Fuel feed: A minimum clearance of 18" shall be maintained from combustible materials.

4.7.2 Combustion base: A minimum clearance of 4″ shall be maintained from combustible materials.

4.7.3 Barrel:

4.7.3.1 A minimum clearance of 18″ shall be maintained from all walls without a heat shield. A minimum clearance of 12″ shall be maintained from all walls with a heat shield including 1″ air gap. *[Author comment: This is based on experience in moderate climates such as Portland; for very cold climates, where the heater may be run longer and have stronger chimney draft, things may get hotter. Consider heat shielding and clearances around the metal barrel as locally approved for wood burning stoves.]*

4.7.3.2 A minimum clearance of 36″ shall be maintained from ceiling.

4.7.4 Heat exchanger:

4.7.4.1 Minimum distance between ducting and combustible wall shall be 6″.

4.7.4.2 Any fabric used on seating surfaces shall be heat tolerant.

Inspection Points:

An inspection should be performed after ducting has been installed and while it is still exposed.

Confirm measurements:

1. Proper ratio of heat riser to burn tunnel and heat riser.
    Heat riser is twice the length of the burn tunnel.
    Heat riser is three times the height of the fuel feed.
2. Confirm proper cross sectional areas.
    Consistent cross sectional area throughout system except barrel, manifold, and possible 30–40% decrease in final third of flue length.
3. Confirm clearances.
    18″ from fuel feed to combustibles.
    18″ from barrel to walls without heat shields.
    12″ from barrel to walls with heat shields with a 1″ air gap.
    36″ from barrel to ceiling.
    4″ from combustion base to all combustibles.
    Room for sufficient masonry thickness around ducting.
4. Confirm suitable footing and foundation.
5. Confirm appropriate cleanouts.
6. Confirm presence of flue priming port or other means to ensure proper drafting.
7. Confirm exhaust flue is protected from precipitation and vermin.
8. If pressure controlled mechanical ventilation is not present, confirm exterior air supply.
    Inlet is located outside dwelling or in non-mechanical ventilated space.
    Inlet is protected against precipitation and vermin.
    Outlet is located in the same room.
    Outlet is located below fuel feed.
    Air supply has shutoff valve.
9. Confirm existence and orderliness of maintenance and operation manual.

Appendix 3:
Special Cases

The following projects represent promising or successful experiments that address common questions and interesting problems that many readers may be interested in solving. They suggest the wide range of possible applications of this technology.

Alternative Chimneys

Horizontal Exhausts: Not Recommended

We no longer recommend through-wall or horizontal exhausts for most installations. See sidebar "Chimney Methods: Up or Sideways?" in Chapter 4.

Many people have the optimistic impression that a horizontal exhaust may improve the cost, ease of installation, efficiency, or discretion of the project. However, because they work so unreliably, ongoing attempts to remedy the initial mistake can end up costing more and requiring a lot more awkward work — and these chimneys usually continue to perform worse than a vertical, through-roof chimney.

Some people still want use horizontal exits, either as a dogleg through-wall chimney, or a customized vent. There have been occasional success stories when most of the following conditions are met:

- Low-height buildings: typically under 10 feet. 16 to 20 feet (1 story plus attic) is about the tallest we've heard of a successful side-exit without mechanical boosters, and many of these are eventually extended into a full-height chimney above the roof.
- Buildings with membrane roofs/no roof vents to compete with a wimpy exhaust chimney.
- Mild climates where outside chimneys don't chill too much (and a non-functioning heater won't be a life-threatening problem).
- Damp or very cool spaces, such as a greenhouse (downward, sheltered exhaust).
- Situations where a chimney might be the only thing that stops the project going forward — for reasons such as cost, shared tenancy, or peculiar local laws or

covenants. *Caution: a project that doesn't work can set back local acceptance even further.*
- Situations where there is some reason to try for partial success now in order to save up for a proper chimney later, such as splitting up the cost of the project.

SIDE EXHAUST DOs:

- Exit high, preferably as close to the ridge height as possible (on a gable wall or toward the higher end of a shed roof), to get the benefits of an almost-vertical chimney. Stay away from dripping eaves, especially in snowy/icy climates.
- Leave yourself options to do the best possible chimney: to extend to above the roof, build an insulated enclosure, or even replace the experimental exhaust with one inside the building. For example, you could build a capped T into the bench as a takeoff point for plan B.
- Screen. The visitors found in unscreened rocket stoves to date include rats, packrats, mice, skunks, bandicoots, frogs, newts, insects, snakes, songbirds, and a duck.
- Weatherize: Use a sealed-on collar to divert water running down the chimney before it hits the wall. Consider a drain port at the bottom of the exposed chimney.
- Fire-proof and smoke-proof: Don't exhaust near air intakes, windows, doors, or right under roof eaves. Use high-temperature construction methods.
- Limit drifts: Keep fly ash, snow, leaves, dirt, and debris from blocking the exhaust.
- Control wind: Make the wind flow *across* the exhaust outlet, not into it, no matter what. Wind gusts, pressure, and building eddies (such as wind whipping around the eaves) can push exhaust backward down the exit pipe. Tools include:
  - pivoting "weathervane" chimney caps for short or cold vertical exhausts
  - H-shaped pipes made from 3 T's (as seen on boat stoves)
  - wire cages of large rock (like a country fencepost or gabion)
  - barrels with lots of ventilation holes
- DO EVERYTHING ELSE RIGHT. A system with any other marginal elements will compound the draft problems that are likely with an unconventional chimney.
- See "Chimneys," in Appendix 4, "Dedicated Outside Air" and "Make-up Air" below, and "Chimney Methods Up or Sideways?," Chapter 4.
- Consider an exhaust fan only if you have other heat sources that will be available in a power failure.

SIDE EXHAUST DON'Ts:

- Don't use an unscreened opening.
- Don't let undersized vents or clogged screens choke the flow.
- Don't expect a horizontal exhaust to out-compete a tall, warm house.
- Don't expect the exhaust to cooperate. Air follows pressure, not your willpower or intention.
- Don't make a "wind scoop" when you are trying to block the wind. For example, a sideways chimney cap can scoop up wind if the wind blows along the exit pipe, then is deflected by the cap.
- Don't confuse poor draft with efficiency. A balanced stove self-regulates the draft needed for clean, efficient combustion at the firebox. A stove that doesn't work reliably can't be considered efficient, because

it is useless when it is needed most. When we don't fully "maximize" a particular factor, such as low exhaust temperature as a proxy for efficiency, it's because we feel a functional compromise gives better results.

Using an Existing Masonry Chimney

Existing masonry chimneys have been used successfully for many rocket mass heaters, but as their conditions vary widely, we did not discuss them in Chapter 4. Appropriate flue sizes are listed in the "Equivalent CSA for Gas Flows" sidebar in Chapter 6.

Terms

A flue is a lined path for smoke or exhaust. Each flue or pair of lined flues has its own structural support, such as a concrete chimney block, a full width of brick between each pair of tile-lined flues, or a separate ceiling box and through-roof kit for each manufactured (metal) chimney flue.

A chimney is a structure which may contain one or several flues.

A port is the term we use here for any opening into a flue: the port may include a cleanout access door, attachment collar, thimble for connecting stovepipe to masonry chimney flues, or other hardware.

No Shared Flues

Two combustion devices should never share one flue. It is not possible for a flue to be the right size for each device operating alone, and have sufficient capacity to safely exhaust both devices when operated simultaneously. In the case of rocket mass heaters, the low-temperature exhaust could be particularly prone to backdrafting downward and escaping into the room through the connected, unused device.

CONSIDERATIONS FOR SUCCESS

Size: The existing chimney should have a lined flue in about the same size as the heater, as described in Chapter 6, "Equivalent CSA for Gas Flows" sidebar. So, 50 square inches for an 8″ heater; 30 square inches for a 6″ heater.

Any ports from room into flue should also be the appropriate flue size: no necking down to squeeze an 8″ heater through a 6″ port. A larger, rectangular hole can be cut in the side of most rectangular chimneys if needed. Cleanout ports need not be the full dimension, just large enough for convenient removal of ash/soot/stray wildlife.

Flues that are too large can be lined with an appropriate (non-corrugated) chimney liner, ideally with insulation between the liner and existing masonry chimney.

Chimney inspection: Any chimney should be inspected prior to being returned to use. Chimneys may be unsafe to use if they have air/smoke leaks, badly cracked or missing liners, creosote buildup, damaged masonry, or obstructions like inoperable dampers or wildlife nests.

Spark arrestors, and caps: Necessary more to exclude weather, vermin, and debris than for sparks, but we do recommend installing them.

Chimney top dampers: We generally do not use chimney top dampers with rocket mass heaters, as the primary air control at the feed is sufficient for most needs. (The heaters do not draft excessively when cool.)

If a chimney top damper is already in place, and does not require modification in order to fit a new liner, it can be retained. Make sure you know how to operate the damper, and make a note in the heater manual or another obvious place. Rocket mass heaters should always be run with any upper dampers 100% open. It may be an even better idea to remove any chimney dampers or throat dampers from a previous fireplace or woodstove.

The optimal masonry chimney for use with a rocket mass heater is:

- Indoors, or well-insulated between exhaust liner and cold masonry walls. Ideally, the masonry can count as part of the heat storage for the heater, and can also help store ambient warmth within the home. Indoor chimneys do this without much modification, if the home is kept warm throughout the cold season. Too much cold masonry, such as for underused heaters or exposed exterior chimneys, can chill the exhaust below acceptable temperatures for successful draft.
- The proper size for the heater (or a suitable heater can be chosen for compatible size with an existing chimney).
- Accessible through a port or opening of the proper size for the heater.
- Not currently in use by any other appliance, unless the appliance will be removed to build the heater.
- Lined with tile or stainless steel liners, in good condition, with insulative backing. Smooth, well-built masonry chimneys may also be used, but should be inspected for good mortar seals. The old method of plastering the inside of a chimney is not very practical for 6" or 8" flues, but could be considered for new masonry chimneys or very large mass heaters.
- Screened and capped in a way that provides excellent flow for exhaust (no tiny screens or undersized openings between screen and cap).

Alternative Air Supply
Dedicated Outside Air

Outside air is dedicated combustion air from outside, separate from household room air.

Outside air was made mandatory in some previous versions of the IRC building code, in some local amendments to the masonry heater codes (R1002), and may be legally required for certain over-sealed homes susceptible to negative pressure, such as manufactured homes. We generally don't like the idea (see discussion in Chapter 6), and current versions of the IRC have dropped the requirement due to complaints of safety problems (it's theoretically appealing, but it turns out buildings are complicated. It is not always possible to safely install outside air in all situations, nor does it reliably fix the problems it was intended to fix). If officials force the issue, you might find a solution in the example we've included below.

One of the big problems with requiring outside air to feed the heater is that air cannot enter at the bottom of the firebox because that would break the thermosiphon and allow smoke to rise up the fuel feed. And air cannot enter much above the level of the fuel feed, or from an upward direction, because that could cause the outside air intake to likely act as a secondary chimney, drawing the fire backward (a known problem with outside air in all types of stoves). So the only way to do this is to get the air to enter the firebox roughly at the same elevation as air enters the normal feed opening.

The outside air should rise to this level from below, ideally from below the level of the burn tunnel floor.

Exclusion of House Pressures

True outside air requires separation from room air (otherwise, it's basically just "make-up air" — which we vastly prefer [see below]). Any door or box built around the fuel feed to separate outside air from room air must:

- tolerate the intense radiant heat from coals in the firebox
- be configured so that it can be opened for loading when hot, and so that the resulting hot surface does not draft smoke upwards into room
- align with the fuel feed for convenient loading, and allow ash removal
- allow control of air by separating incoming outside (or room) air from the firebox at the usual location of the feed lid
- allow viewing and tending the fire (glass or at least translucent mica).

Make-up Air

Exterior Air for Room Ventilation

Make-up air is fresh air delivered into the room, to "make up" for air being removed by appliances and house ventilation. (Air for this purpose can also be provided as part of a balanced ventilation system for the house as a whole, eliminating negative pressure problems at the larger scale; when the home has balanced ventilation, it may no longer need "make-up air" because the problem is already solved.)

In general, make-up air is much easier to work with than dedicated combustion air, and a better long-term solution for efficient, safe heating and general indoor air quality.

There are a lot of easy ways to supply make-up air safely, such as heat-recovery ventilators (a box of tubes in a windowsill), a screened hole in the wall or crawl space, or even just opening a window if there are too many appliances running exhaust fans at the same time.

Make-up air can be delivered in any convenient location in the room, well away from fire and fire dangers. It can even be delivered from above, though this does less to alleviate negative pressure problems. The greatest negative pressures in a building are at the bottom, so the lower an air inlet is, the more consistently it is likely to draw air inward.

The main trick with make-up air is avoiding cold drafts. Heat exchangers to preheat incoming fresh air are much easier to manage than dedicated firebox air, with fewer risks and fewer practical constraints. Windowboxes, or separate, fresh-air channels behind the heater are two workable options.

We have also seen designs for pipes embedded in walls, designed to preheat cool outdoor air as it rises up the wall before entering the room.

Air Requirements for the House Itself

Any home experiencing negative pressure problems should be checked for general healthy air flow. Over-sealed homes often develop negative pressure, as well as poor indoor air quality, which can be alleviated by a good fresh air supply. The minimum healthy ventilation recommendation from one Seattle energy audit provider was one third of the home's volume exchanged per hour.

Outside Air Prototype:

(Multiexperiment prototype with 8" flues and firebox, pebblestyle mass, cobbish firebox core.)

Construction Notes:

- Air feed: 5"x10" rectangular duct, same CSA as 8" system flues (50 square inches).
- Air enters the building near floor level, from 2" to 12" above floor.
- Cowling or "bubble": Constructed of brick with clay sand mortar: Interior about 11" by 21" by 9" tall. Wide enough for air control firebricks to slide completely out of the way for fire handling. Large size also reduces radiant heat exposure (a tall, hot, narrow feed draws smoke upward).
- Cover: A piece of reclaimed stove door glass with woven fiberglass gasket.
- Air control: Bricks at feed opening give normal range of control while running on either outside air or room air. Outside air can be blocked, if needed, by covering the duct where it opens into the cowling.
- Safety: Sparks can pop or drop into the air feed, and weird wind pressures could drive flames and smoke backwards into it. Outside air ducts must be noncombustible. Protect nearby combustibles from the potential heat (with air gaps or noncombustible insulation).
- Results: Worked as intended, even provided some extra draft (maybe due to colder air falling down the feed helping the thermosiphon). However, the metal got very cold. The owner says "The rocket mass heater uses hardly any wood ... and hardly any air. But when the rocket mass heater is running, the air in the room is fresher."
- Current status: Decommissioned; the wall inlet was blocked and insulated.

Photos courtesy Paul Wheaton, originally shared on www.permies.com

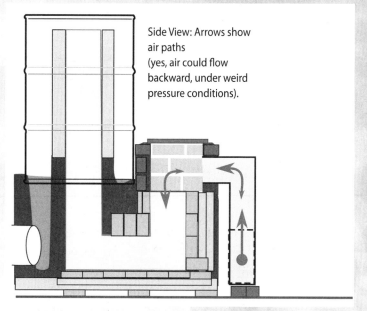

Side View: Arrows show air paths
(yes, air could flow backward, under weird pressure conditions).

APPENDIX 3: SPECIAL CASES

Top view:
Air comes from outside at lower right, rises to the brick cowling or "bubble" around the feed, then pours toward the fire. When the air and cowling are cold, any smoke puffs to sink back into firebox. Our brick sides are deliberately big, out of the main radiant heat from the fire. A hot cowling or too tall feed opening can draw smoke upwards from the fire. Since the glass lid must be opened to feed the fire, any smoke collecting under the lid could be a problem. The owner learned to crack the lid and check for smoke before opening it fully.

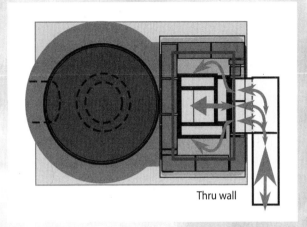

Thru wall

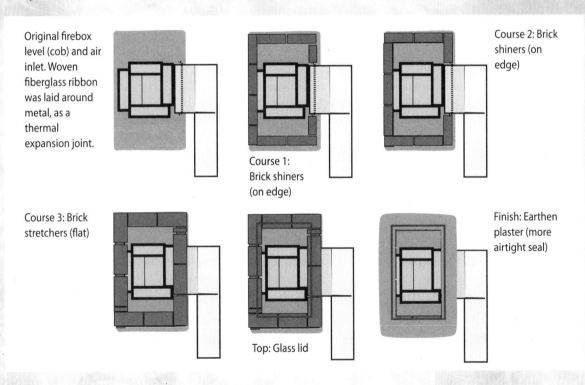

Original firebox level (cob) and air inlet. Woven fiberglass ribbon was laid around metal, as a thermal expansion joint.

Course 1: Brick shiners (on edge)

Course 2: Brick shiners (on edge)

Course 3: Brick stretchers (flat)

Top: Glass lid

Finish: Earthen plaster (more airtight seal)

Example Solution: Wheaton Labs Office Heater Experiment

Here we show one example where outside air was added to a rocket mass heater prototype, while avoiding many of the problems above. We have heard of a few other examples (not shown) where outside air was successfully installed by ducting it directly into the metal feed cowling shown in Evans/Jackson book *Rocket Mass Heaters*.

Alternative Thermal Mass

The heaters described in this *Builder's Guide* all share an all-masonry thermal mass. We often receive questions about alternatives, specifically, water or radiant fluids as thermal mass (e.g., for aquaponics or as a substitute for a boiler), and "lighter" thermal mass or removable mass (gravel, dirt, sand, etc., usually encased in a wooden box for "temporary" or "portable" installations).

Water As Thermal Mass

Water has excellent heat capacity, but it has other characteristics that make its performance very different from masonry heat storage, and potentially more dangerous. This is not a complete guide to the issues or safe design protocols — just a general primer that may be useful when evaluating other people's designs or when assessing the risks and success chances of developing such a design.

Explosions: Water boils at 212°F, expanding about 1700 times in volume to become steam. If this steam expansion is trapped in any container, pressurized coil, or clogged pipe, the result can be explosive. Any closed or pressurized water-heating system must be well designed to minimize the risk of explosion, and even well-designed systems can fail with lethal consequences. Training and experience are recommended.

Incompatible safe temperatures for fire and water: Water boils at 212°F, and clean fire burns between 1000 and 2000°F (roughly 500 to 1200°C). This means that an exposed water jacket or coil cools any flames it touches below clean burn temperatures, causing creosote deposition (and risk of future extremely high-temp fires, for example if the water coil gets blocked or emptied). It can also flash the water to steam, causing risk of explosion even in non-pressurized systems (such as open thermosiphoning pipes), if the pipes reach sufficiently high temperatures. Wood heat is variable and hard to control compared with refined fuels or even sunlight.

Life and health safety: Hot water storage involves a choice between scalding hazards and breeding pathogenic microbes in lukewarm stagnant water.

Water as universal solvent: Minerals, salts, and life-forms will move along with the water, and deposit as gunk in the system. This is even more true in open systems, which use openings to reduce the risk of overpressure and explosions. Damp escaping from water heaters can accelerate corrosion and rot of other nearby materials, such as structural supports, metal parts, wall surfaces, textiles, etc.

Flooding: Leaks or safety venting can cause flooding and damage. All tanks fail eventually, unless they are retired first. Auto-filling tanks with a major leak can release a far larger volume of water than the tank itself can hold.

Freezing: Any plumbing system includes the risk of damage when water freezes (usually 32°F/0°C, but varies with altitude and salinity). Even if the pipes don't burst, re-heating a frozen system carries increased risk of overheating at other points, as water will not flow until all areas are thawed.

Other fluids: Some systems use other fluids, such as steam, propylene glycol, oils, or antifreeze to improve performance or limit certain problems. Each fluid has its own properties and risks.

Heat capacity and load patterns: Heating water takes a lot of energy. Hot water is used in batches, year-round (contrasted with home heating, which is seasonal and steady in most climates). It may be better to have a dedicated hot water heater, and a dedicated household heater, than to try to get both functions out of the same fuel load while adjusting for seasonal and daily changes in demand.

Successful Water-heater Examples

Erica's personal wood-fired water-heating is very simple: a tea kettle or an open stockpot with tap that can be drained and removed for cleaning and inspection. These can be operated safely by most people with minimal risk, though they don't necessarily deliver the amounts of hot water needed for large-volume heating or bathing.

ZAYTUNA FARM WATER HEATER

Geoff Lawton explains the basic design in this video: http://permaculturenews.org/2013/05/20/how-to-build-a-rocket-stove-mass-water-heater/.

Installer Tim Barker shows the construction process and answers technical questions in this article: http://permaculturenews.org/2012/11/23/rocket-stove-hot-water/.

A lovely system at Zaytuna Farm in New Zealand uses a large water tank inside an insulated steel cavity to collect heat from the rocket flames and exhaust. The large water tank is plumbed with steam outlet, inlet valve, and a secondary coil that carries domestic water through the tank. The potable water circulates briefly through the coil, at gravity-fed pressure of 15–30 lbs. The direct-heated tank can safely boil if overheated because steam escapes through the

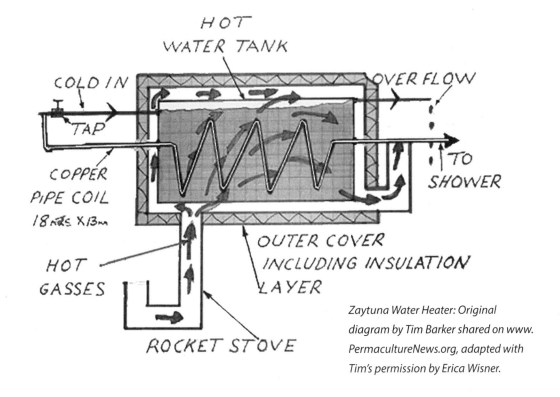

Zaytuna Water Heater: Original diagram by Tim Barker shared on www.PermacultureNews.org, adapted with Tim's permission by Erica Wisner.

large steam outlet which is always open. The pressurized, potable water is never exposed to flame or to risk of boiling (unless the tank completely empties). The system becomes less efficient when the water level drops, so even novice operators are likely to notice the difference if they forget the weekly top-up. A similar system Tim built at a different site uses a float valve to maintain its proper level without human intervention. Students have run the Zaytuna system with notable success for several years.

Tim Barker is working on a booklet that covers this water heater and other useful rocket projects.

Ernie has been involved with the construction or design of several water-heating prototypes, including designs with a heating coil that circulates water instead of a directly heated tank. Some systems use thermosiphoning (gravity- and heat-induced flow) to allow the system to work without reliable electrical power. Successful design and maintenance of such systems requires some experience and fine-tuning. Other experimenters prefer active pumps and monitors to manage flow, pressure, safety factors, and performance.

All piped hot water systems should be considered as potentially pressurized — and dangerous. Design, construction, and maintenance should involve someone experienced in hot water systems, boilers, steam fitting, or other high-temperature plumbing fields.

Many solar hot water systems and other relevant resources are discussed here: www.builditsolar.com, and steam and radiant boiler systems in general are discussed by knowledgeable people on the forums at www.heaterhelp.com, and by creative amateurs and farmers at www.permies.com.

"Portable" Mass Heaters

This example was almost included in Chapter 3, due to its initial success, but the owner only operated it occasionally for a season or two before changing homes. We therefore do not have reliable data about performance, and slow heat damage would not yet be apparent.

Ely 8": "Portable" Box-and-fill Heater

BUILDING CONSIDERATIONS

The A-frame cabin is essentially a wood-framed tent on stilts, in a gorgeous Minnesota birch forest. The floor was never designed with masonry in mind, nor the whole for long-term heating. The roof/walls were not yet insulated, for example, which is normally a high priority for any heated space.

DESIGN CONSIDERATIONS

The owner wanted to gradually improve the cabin into a four-season retreat. This project replaced the gable-end woodstove with a heater that holds warmth overnight. Because a weekend cabin sees a lot more cold starts than warm ones, we recommended including a bypass to preheat the chimney for easier starting.

Rather than cut into the floor joists and build a masonry footing, the owner shored up the floor with extra timber and piers, and built a lower-mass, gravel-filled box for the heater. As with many small cabins, the bench needed to tuck around existing walkways and windows, and serve as built-in furniture. The owner was very happy with the resulting L-shaped bench, wide enough to double as a warm cot for guests.

Ely 8" Heater: photos courtesy of project owner.

Site Details

Project dates: 2012
Location: Northern Minnesota
Building size: 800 sf
Chimney height: 20 ft
Foundations: Timber on piers (8˝ x 8˝ posts, 4 feet on center, with diagonal bracing)

ELY 8" Rocket Mass Heater

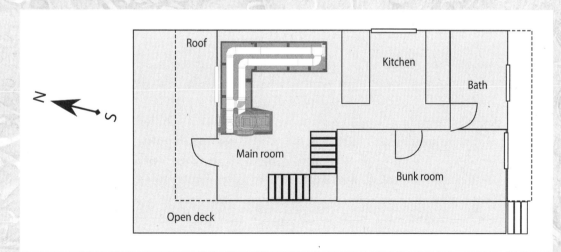

Other Portable Heaters

Paul Wheaton is probably the best-known proponent and developer of "portable" rocket mass heaters, a variety of demonstration or temporary-use system that is specifically tailored for eventual removal and reconstruction elsewhere. Several successful experiments have been done by Paul and others, ducting the exhaust from a rocket combustion unit through a wooden box full of sand, gravel, or other mineral mass. (See ongoing updates at http://www.permies.com/t/2558/rocket-stoves/portable-rocket-mass-heater.)

Useful tips for this type of mass heater include:

- THEY ARE DANGEROUS: Incorporating wood around or under a heater takes us well outside accepted best practice. We have had prototypes catch fire and cause incandescent concrete spalling (popping red-hot rock chips out of a damp pad); we've also discovered undetected charring on heaters when they were deconstructed. These designs are not fully developed, and therefore not safe. While several have been built and used with some success, the parameters for safe design and long-term use are not well understood.
- Combustion unit must be designed for safe clearances from any woodwork, and/or for a robust connection with the mass if the combustion unit is built separately. We prefer at least 4″ of mass around the heat-exchange ducting, and substantially more (with multiple layers of refractory insulation, and no aligned seams) in the combustion area. We have had some good results using metal "bulkheads" to bridge too-hot areas of wooden box heaters.
- Dirt seems to give better performance than gravel, especially dirt that's moistened and tamped in during filling. Gravel does better than sand or large rock chunks. Pea gravel or ¼″-minus road grade, with any large rocks you happen to have lying around, also seem to work reasonably well.
- Mass is still heavy. Strap or brace any box walls to withstand lateral (outward) force, and reinforce floor to handle expected weights of both mass and people enjoying its warmth.
- People are monkeys. Cats and dogs are people too. Uncovered loose mass is liable to get played with, used as a litterbox, etc.

Our primary concerns with portable mass heaters are:

- **Flammability and fire hazards:** The common use of wood or other combustible materials in the framing for the mass containers means that surface temperatures must be carefully monitored. These heaters don't qualify as masonry heaters under any code, nor are they likely to. Our favorite version of this approach is probably a masonry box such as brick or CMU (concrete masonry units, e.g., 4″ cinderblocks or pavers), which has the benefits of portability without the drawbacks of combustibility.
- **Exhaust leaks:** Loose mass circulates room air. Most "portable" heaters are made with a single layer of stovepipe (some owners upgrade to seamless or welded stainless pipe). Any pinhole leaks, damage, or disconnected pipes could easily dump exhaust fumes into the room air without much chance of detection. Combustion

exhaust is never safe nor is it desirable to breathe. Even if the heater typically maintains a visually clean fire, there may be traces of smoke or CO that can poison occupants. Strategies tried so far include using welded pipe, sealing joints with high-heat tape or silicone, careful building and support, and use of CO monitors and smoke detectors to warn of leaks before disorientation due to CO poisoning can set in.

- **Efficiency:** Most loose-mass heaters do not store heat as effectively as solid mass heaters. In one collaborative comparison, a pea-gravel-and-rock heater lost about 20 degrees F overnight, while a similar sized, solid, rock-and-earthen-mortar heater lost 10 degrees F or less overnight. In areas not used overnight, converting more heat to warm air during firing may be an acceptable tradeoff.
- **Weight:** If the goal is to reduce the overall weight, using less-conductive mass (like gravel with air pockets) means the mass does not absorb as much heat, and more mass may be required in order to achieve the same heating potential.

A smaller, dense mass (such as a soapstone heater) might give better performance for weight than a large, loose mass. Kiko Denzer's Masonry Heater Hat over a small woodstove is one example (handprintpress.com); a small mass at higher temperature can store as much heat as a larger mass at lower temperature, though it also loses heat faster. There has been some discussion of building heaters with removable mass, such as insetting a small tank or pot(s) of water that can be drained for transport.

Box-and-fill construction can also be done with noncombustible boxes (masonry board, brick, cinderblock, stone slabs in a steel frame, etc.), in which case the project can be much closer to masonry-heater performance and safety practices. Benefits and appropriate situations where a box-and-fill heater might be useful:

- **Alternative aesthetic appeal:** Different from curvy cob or traditional masonry. Portable heater demonstrations have gotten many people to consider mass heating, and even to appreciate the benefits of a more permanent masonry heater, despite initial lack of interest in the larger masonry systems.
- **Demonstration projects and prototypes:** Great for proof of concept at events, for class groups, and perhaps for skeptical authorities (it might just convince them you're really crazy). For prototyping, loose-fill mass heaters can only offer a first approximation of the performance of a solid mass heater. But it's a better approximation than working strictly from theory.
- **Resource, skill, and time limits:** If suitable earthen building soils are not available, a box-and-fill method can allow heating using any mineral soil. We prefer a masonry box for safety (brick or concrete or cement-board). Box construction takes time, and overall the process is not necessarily faster for a skilled builder. But it might allow a good welder or cabinetmaker to stick with familiar skills. Filling the box, once built, takes far less time than site-built masonry.
- **Damp conditions:** We recommend masonry box-and-fill systems for raised

greenhouse beds, where a cob or earthen masonry structure would be subject to constant damp, and where being able to move the heater or adjust heights and depths could be a big asset to crop production.

- **Tenuous tenure and transport:** If the fill is not transported, a box-and-fill system might allow nomadic folks to transport their heater in a camper, manufactured home, etc. New local fill could be added at each location, avoiding the cost of transporting the heavy mass.

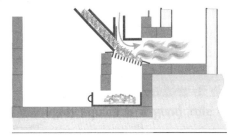

Pellet hopper: photos and description courtesy of Rob Torcellini, originally shared on www.permies.com and YouTube.

Work is ongoing to develop a "shippable core," a combustion unit that could be installed in one or a few pieces, to facilitate very fast installation of either portable or standard mass heaters. It turns out this is a very hard set of functions to combine, especially while keeping it affordable.

Alternative Fuels
Pellet Fuels

Rob Torcellini of Bigelow Brooks Farm posted a report on his successful pellet-hopper prototype for burning wood pellets in a rocket mass heater.

There's discussion and further developments on the forums at www.Permies.com: http://www.permies.com/t/18515/stoves/Burning-Pellets-Rocket-Mass-Heater#163683.

A video, "Burning Pellets in a Rocket Mass Heater," posted by web4deb, is here: http://www.youtube.com/watch?v=lx_9cvd5sQ0.

The following project summary is compiled from Rob's video and online remarks as of 10/26/2012:

> I had already built a standard rocket mass heater to take firewood, and used it for about a year to heat my greenhouse and aquaponics. (There's a video of that one on YouTube, too.)
>
> I wanted to try burning pellets instead of firewood. Some pellet stoves use an auger that's driven by electricity, but I wanted to try just a gravity feed first.
>
> I pulled some of the brick away from the front of the wood feed to make room.

I welded up a grill tight enough to hold the pellets up above the ash, and started prototyping hoppers made of angled steel tubing. The first version was not giving me enough heat — there wasn't that rockety sound. The opening was too small, and the pellets were not spreading out across the grate. And with the vertical hopper orientation, it got way too hot and the fire started burning back up the hopper.

My next version is made from a wider C-channel that spreads the pellets better. Across the opening, I added a sliding gate to control the feed rate. I also made an air restrictor so that air is forced across the pellets, and another air gap under the ash pit that draws air up through the grate. The rockety effect is way better; the air speed forces better mixing and a hotter burn.

The pellets usually feed in in clumps, so if you look through the air gap it seems like they're smoldering a bit. But that's the initial gasification; that smoke gets burned further along in the burn tunnel (and heat riser).

The steel grate may burn out in the intense heat, so I may eventually need to try stainless. The steel heat riser is completely burning out [shows flakes of rusty steel fragments]; I used a galvanized pipe and it didn't last long. For now, I'm hoping the mineral wool is holding itself in place without the steel liner, cause the steel is definitely gone. ... When I dismantle the system, I'll do an alien autopsy video on it. Then I'll assess how to repair/replace it.

It's burning around 8 lbs of pellets per hour, which is approximately 64,000 BTU's per hour. It gets a lot hotter than it did when I was burning wood. The barrel used to max out around 500 degrees F up near the top, and now it's 500 F halfway down and about 300 F at the base.

Update from Rob on 11/28/2012:

The pellet feeder is working awesome! Yesterday I did my first long-term burn in it (over 2 hours). I burned 80 pounds of pellets in 13 hours. I'm not sure if you're familiar with the greenhouse setup, but I'm heating the greenhouse with an aquaponic system in it. I have about 1200 sq feet of greenhouse building and it has around 2000 gallons of water. I burned yesterday since it was snowing and I had basically no solar gain. I was able to maintain the building temperature at 60 degrees (30 outside), and heated the water from 55 to 63. (My fish were much happier in the warmer water.)

The only problem that I've had is that some of the pellets don't burn completely and they start to accumulate on the grate and don't fall through into the ash pan so it starts to clog it up a bit. I think there's some foreign matter in these pellets (rock, metal?). This started to prevent the ash from falling through and would slow down the burn a bit. About 8 hours into the burn, I quickly scraped the grate a bit to clean it off. Considering the amount of maintenance I had to do during the burn, I would call it a success! That was about $8.50 in pellets from Home Depot. For me, the time (and my back) I saved buy not splitting is well worth it! Next up, researching hammer/pellet mills!

This beast is an inferno when it's running. I was expecting it when I originally built it, but the steel grate is about 50% of the original thickness so I'll probably make a new one out of stainless. It will last longer, but even 302 stainless isn't rated much more than 1500–1700 degrees before it starts to flake. It's also sort of strange to see some of the clay bricks glow.

Rob did a great job explaining not just what's working, but how he problem-solved the initial designs until he got there. He gets a lot of well-deserved compliments on his work.

As Rob notes, there are a couple of problems with using metal around these high temperatures. The temperatures in there can get up around 2400°F just with wood, and with his pellets heating the barrel about 25% hotter, I'd guess the heat riser is hotter too. I'd definitely go for refractory materials like brick or even clay in preference to metal right in contact with insulation batt.

The greenhouse is a great setting for this kind of experimentation: big heat loads, yet forgiving if you have to stop for a few days and rethink. It also has huge vents in case you make smoke. A backfire up the hopper could be tragic in a home, and not real pleasant even in the greenhouse.

I liked that Rob's first hopper design was the same height as the burn tunnel, even though it didn't work that well — it was a safer experiment to start with, and then he worked from what he learned before going to the taller version he uses now.

You'll note he's got plenty of space around it even now that it's working well. And it turns out that was a good idea — as of 2015, there's a further update reporting that the pellet hopper gummed up with wood resins and the whole hopper load caught on fire, making a seriously terrifying reverse rocket stove. Rumor has it Rob is considering batch-fed options rather than continuous feed for his next experiment.

We extend our thanks to Rob for sharing his story. If you want to support Rob's work or see more of his projects, you might wander over to www.BigelowBrook.com/donate.

Coal and Petroleum Fuels

At least three coal-burning rocket mass heaters have been successfully built at the time of this writing; we have two reports of successful examples in Mongolia. (We are interested in diesel or trash burners, but don't have a good prototype to share yet.)

To burn coal and other fossil fuels safely, there are several considerations:

- Air supply: Petroleum fuels are very rich, and they require proportionately more air to burn successfully. They will also maintain a fire with excess air more readily than wood, due to their greater heat production. Make air feeds larger and provide controls to adjust the air as needed. (One of the Mongolian experiments combined this larger air feed with a wok-stand over the air intake for high-heat cooking ... fascinating!)

- Fuel density: Pound for pound, coal has about twice the fuel value as wood. This means less weight to haul and less dry storage needed. Coal may be a locally accessible fuel in areas where deforestation is a major problem and wood fuels are scarce.
- Heat of combustion: Coal burns hotter than wood, especially when compared with common, slightly damp fuel wood. Interior liners should definitely be firebrick or similar high-temperature refractory materials (rated for 2400°F or better; 2800°F would be reassuring).
- Fuel feed: Coal comes in lumps rather than sticks, so it doesn't self-feed the same way under gravity, and the air will not be forced between the fuel chunks as with wood. A grate that raises the coals an inch or so off the bottom of the firebox floor may be useful. A horizontal feed opening (like a fireplace or woodstove) instead of the downfeed J-tube may be worth considering, as the J-tube's advantages are more relevant with stick-shaped fuels.
- Sustained flame: It can be more difficult to light a coal fire for those unaccustomed; some people start with a wood fire then add coal once it's hot.
- Chemical pollution: Some kinds of coal contain high amounts of sulfur and other contaminants that will generate corrosive sulfuric acid to damage your stack and your neighborhood. Coal may also produce carbon monoxide with less detectable scent than wood smoke; do not let clear exhaust deceive you into trusting it.

Unfortunately we don't yet have pictures of the coal-fired rockets.

Other Fuels

Other "special" fuels that are mostly cellulose can be burned about the same as wood, if you can work out a tolerable feeding schedule and spark-safe storage. These include straw, sawdust, dung, plain cardboard, shrubs, dung briquettes (from horses, goats, sheep, cattle, buffalo, yaks, donkeys, camels, grazing elephants — but *not* omnivores like pigs, dogs, or people, nor carnivores).

Dried bundles of invasive weeds, orchard prunings, and coppiced wood are also extremely efficient low-cost or free fuels—that have additional benefit of improving the local environment.

NEVER BURN WET FUEL. Keep all fuels stored as dry as possible. (See Chapter 5 for details.)

Extreme Sizes

Peter van den Berg's 8" Batch Box Beastie, AKA "The Thing"

Can I make a bigger rocket mass heater for my large space?

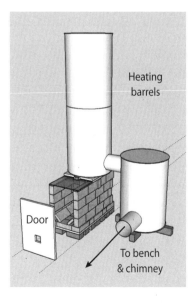
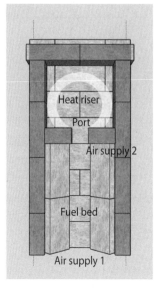

Batch Box: Diagrams and tables courtesy of Peter van den Berg, originally shared at donkey32.proboards.net and www.permies.com

Can I make a big firebox that loads from the side like my woodstove?

Peter van den Berg from the Netherlands has an answer. His batch box builds on the big fireboxes designed by Lasse Holmes of Homer, Alaska, and optimizes for very clean combustion. These designs combine the insulated, clean-burning chimney of a rocket mass heater with a larger firebox. By passing the gases from the big box through a narrow gap, or "port," Peter's design induces a double vortex which he calls the "double ram's horns," whipping the air and fuel gases for an amazingly clean burn.

This design requires precise proportions, and some extra elements such as a secondary air feed (the L-shaped metal tube near the center of the box, visible just below the disc Peter is placing in the photo).

These features are essential for a clean burn, and thus for efficient combustion and

Peterberg Rocket Stove Calculations

Heat riser dimensions inches

Diameter	3	4	5	6	7	8	9	10
Length	21 5/8	28 13/16	36	43 3/16	50 3/8	64 13/16	64 13/16	72
Base (for calculations)	2 3/16	2 7/8	3 5/8	4 5/16	5 1/16	5 3/4	6 1/2	7 3/16
Box dimensions								
Width	4 5/16	5 3/4	7 3/16	8 5/8	10 1/16	11 1/2	12 15/16	14 3/8
Height	8 1/2	8 5/8	10 13/16	12 15/16	15 1/8	17 1/4	19 7/16	21 5/8
Depth	8 5/8	11 1/2	14 3/8	17 1/4	20 3/16	20 3/16	25 15/16	28 13/16
Port dimensions								
Width	1 1/16	1 7/16	1 13/16	2 3/16	2 1/2	2 7/8	3 1/4	3 5/8
Height	4 3/4	5 5/16	7 15/16	9 1/2	11 1/16	12 11/16	14 1/4	15 13/16
Depth	2	2	2	2	2	2	2	2
P-channel								
Width	1 1/16	1 7/16	1 13/16	2 3/16	2 1/2	2 7/8	3 1/4	3 5/8
Height	5/16	7/16	9/16	5/8	13/4	7/8	1	1 1/16

Peterberg Rocket Stove Calculations

Heat riser dimensions millimeters

Diameter	75	100	130	150	180	200	230	250
Length	540	720	936	1080	1296	1440	1656	1800
Base (for calculations)	54	72	94	108	130	144	166	180
Box dimensions								
Width	108	144	187	216	259	288	331	360
Height	162	216	281	324	389	432	497	540
Depth	216	288	374	432	518	576	662	720
Port dimensions								
Width	27	36	47	54	65	72	83	90
Height	119	158	206	238	385	317	364	396
Depth	50	50	50	50	50	50	50	50
P-channel								
Width	27	36	47	54	65	72	83	90
Height	8	11	14	16	20	22	25	27

to avoid loading downstream heat-capture channels with creosote.

Peter estimates this heater's working mass by the internal surface area the exhaust will heat: 9 or 10 square meters for this 8" size. In the demonstration, he achieved this with three barrels and a 30-ft straight bench. We don't know if it requires a larger gap above the heat riser (it had about a meter of gap in Peter's prototype layout), or how it would operate with a very long or convoluted bench. A spreadsheet calculator for the dimensions is available at: http://donkey32.proboards.com/thread/734/peterberg-batch-box-dimensions

We do know from Peter's work that a vertical chimney is essential for these designs. The whole system must operate under-pressure, provided by the exit chimney's draft.

Many thanks to Peter van den Berg for contributing these images, and for posting the complete report at www.permies.com.

Mosquito Heater

We went the other direction, and made an absurdly tiny batch box at the 2015 Innovators' Gathering. The firebox and heat riser are carved out of the insides of two pairs of insulating firebricks, with a diameter of just 2 inches. Not surprisingly, it runs fitfully and tends to smoke. Fire is more stable at larger scales, especially with variable natural fuels like wood.

Insulation doesn't seem to scale, either: if an 8" firebox needs 1" of heat riser insulation, a 2" firebox will also need a full 1" of insulation, or even a little extra, if you want the inside and outside temperatures to be the same.

The smallest practical fireboxes for mass heaters seem to run about a 4" exhaust stack, and they are still very fiddly to light and operate cleanly. Fireboxes smaller than 5" generally do not have enough extra heat to run exhaust through channels in a mass storage bench.

Folks do report success setting up 4" fireboxes as stand-alone cookstoves or on-demand radiant heaters, or with a masonry bell or barrel full of bricks directly around the firebox instead of a channels-through-mass design.

The online forums at donkey32.proboards.com and www.permies.com are great places to find discussion about ongoing developments, especially while the tiny-house craze is still going strong. There are several Facebook groups dedicated to rocket stoves and rocket mass heaters.

A 2" batch-box style heater - burns dirty, but surprisingly, it creates the same vortex patterns as its larger cousins.

Appendix 4:
Home Heating Design Considerations

Heating Load Calculations

WE OFTEN GET ASKED if a rocket mass heater will heat so many square feet. Turns out climate, house design, and personal habits matter more than the floor footprint of the house. But it's not hard to get an estimated heat load from your old heating bills and/or a tape measure.

Please Note: If you do not enjoy math, you do NOT have to calculate your home's heat loss. These equations are *no more accurate* for everyday households than just reading your old heating bills, or guessing based on neighbors' wood piles. If you don't love math, you may just want to skip down to the practical tips for adjusting the results that are discussed further down in "Limiting Heat Loss."

Utility Bills and Anecdotal Heating Comparisons

To compare one home to another, with no math, start by noticing your respective climates. You can use heating degree days (below), or a general heating map such as the one at energystar.gov for insulation recommendations. If your climate is twice as cold, you either need twice the insulation and heat-recovery ventilation, or you need a house about half as big.

We profiled several projects in Chapter 3. The Mediterranean and Bonny projects are from California, where redwood valleys or ponderosa mountainsides are considered chilly. The Daybed and Annex projects were done in rainy Portland, Oregon, where winters are grey and clammy, but 2 inches is considered a lot of snow. The Cabin 8″ and the Ely 8″ and Wheaton Labs projects in Appendix 3 were done in colder climates along the US/Canadian border, where people put on studded tires and know the snowplow guy's name and number. An 8″ heater very similar to all of the above (not shown) was used successfully for some years in Antarctica, where special mechanics are kept on staff to repair the snow-cats when the wind whips their doors off. So really, you can be comfortable with about the same amount of stove in any climate, *as long as your home is modest and weather-appropriate.*

You could estimate your heating load numerically, if you want to, simply by looking at last winter's heating bills. If you use more than one kind of heat (for example a gas furnace plus electric space heaters), you can convert and add them up, or keep them separate. It may be useful to distinguish between your ordinary heating needs and optional hobby or spare room heaters.

BTUs (British thermal units, the amount of heat required to raise 1 pound of water by 1 degree Fahrenheit), "therms" (about 100,000 BTU), or watts (or kilowatt hours, kWh) will be listed on most grid-powered delivery bills. For fuels purchased by the gallon, pound, cubic foot, liter, or cubic meter (oil, coal, propane) the BTU content per unit can be found online, for example at: http://www.exothink.com/Pages/btu.html.

If you need to convert back and forth, for example to add up electric and natural gas heat from different heating bills: One BTU is about 1055 joules; a million BTUs is 1 mBTU or 1.055 megajoules. One therm is about 100,000 BTU. An mBTU is about 293 kilowatt hours; a megajoule is about 278 kWh.

One pound of wood can deliver 6200 to 8700 BTU when burned (if it's reasonably dry, and depending on how much of that heat escapes in smoke/hot exhaust; from http://www.chimneysweeponline.com/howoodbtu.htm. Our wood consumption with a rocket mass heater appears pretty close to the theoretical maximum BTU delivery for wood.

It's hard to predict how much more efficient a thermal mass heater will be than your current heating system — because we don't know how you currently heat, and we don't know how well you'll design, place, and operate your mass heater. On average, when replacing an old woodstove with a rocket mass heater, a lot of owners report they now use about one quarter the fuel wood. Normal results range from half to one tenth the fuel wood. Many owners also report greater comfort.

It's not always easy to predict how a mass heater will compare with forced-air furnaces or conventional utilities. A mass storage heater can significantly reduce the utility bills, *even if it is not run very often,* simply because mass has so much more heat storage capacity than the air commonly used by furnaces and fan-type space heaters. And the warm mass doesn't float out of your house every three hours and require completely re-heating its replacement.

Calculating Heat Loss

Heat loads are more often calculated for large public and commercial buildings that use forced-air heating. These equations may not be accurate for thermal mass heating, nor for single-family homes unless you know your home's construction detailing and weather forecasts with rare accuracy.

If you like math, and equations help you see the relative importance of things, doing the calculations may inspire you to new heights of energy efficiency.

If you just want to run the numbers, go find an online calculator, such as the ones in the reference pages at www.builditsolar.org which conveniently break down the heat loss by sections.

If you want to work it out by hand, without the Internet, you are welcome to follow along with my first attempts:

[(Area1/R-value1) + (Area2/R-value2) …] × (indoor temp. – outdoor temp.) × time= Heat loss (BTU or J)

The resulting units will be BTU if you use imperial R-values, hours, and Fahrenheit degrees. If you use metric (including metric R-values, not given in our tables) and seconds for time, I believe you will get the result in joules. If you don't include the time factor, you have an approximation of instantaneous heat loss at that temperature.

If you want the heater to power through the worst cold snaps, use the lowest temperature you expect to see in your climate. But if you are willing to snuggle up to the heater in a cold snap, or tolerate a dip down to 60 degrees instead of 70, you can probably just look at the average from your coldest month(s), or your average winter low.

We usually predict the heat output based on a 4- to 6-hour evening fire; if you're stuck at home in a winter storm, you could run the heater for 10 hours a day and easily double your heat output. (Unlike automated heaters, this one won't fire up if you're not home. But if someone is home all day, you can run it as long as you like.) In eight years of heating with two different rocket mass heaters, we have only once felt the need to sleep on the mass in an extreme cold snap — 20°F below our typical winter low. With a day or two of extra-long firings, we were back in the bedroom before the end of that cold snap.

Metric/SI Calculations

For metric users, the equation is the same, but the units are different.

Heating Degree Days

To calculate your annual fuel budget, look up the total heating degree days (HDD). Heating degree days are the difference between average indoor temperature (usually 65°F/18°C), and the average outdoor temperature. (When outside temperature is hotter than indoors, those times count toward the cooling degree days, or CDD.) If your average outdoor temperature all year round was 55°F (like a cave), and you heated your home to 65 at the walls, you'd see about 3650 HDD. If your winter averaged 55 and your summer 70, you might see 1800 HDD and 900 CDD.

Florida and California experience something like 2000–4000 HDD per year, coastal areas in Oregon and Virginia about 3000–5000 HDD per year, and cold areas like the Rockies and the inland Canadian border range from 6000 to over 8000 HDD per year. Altitude alone can create differences of more than 3000 HDD per year in a given region.

Find HDD and CDD by state and zip code within the USA at the HUD user portal: http://www.huduser.org/portal/resources/UtilityModel/hdd.html

A search on "HDD + [Country]" will turn up other resources, like these metric HDD data for Europe: http://appsso.eurostat.ec.europa.eu/nui/show.do?dataset=nrg_esdgr_a.

Find the surface areas in square meters, use the scientific-unit R-value or RSI, temperature in C (the degrees difference is exactly the same as in Kelvin, or K, the absolute temperature). The result will be in watts (energy per time) if you don't include the time. For joules (just energy), multiply watts by the time in seconds.

[(Area1/R-value1) + (Area2/R-value2) ...] × (indoor temp. − outdoor temp.) = Heat loss (BTU/hr or W)

R-value to RSI conversion:
The conversion between US/Imperial units of R-value and metric units is:

1 hr·ft^2·°F/BTU = 0.176110 K·m^2/W, or 1 K·m^2/W = 5.678263 hr·ft^2·°F/BTU.

Areas and R-values are the area of a surface, such as the number of square feet of window pane, divided by the insulation R-value for that surface (might be an R-value of 2 for bare windows, or 5 to 10 with insulating blinds and drapes).

Indoor temp — outdoor temp: Use degrees C or degrees K for metric (after subtracting, you'll end up with the same number).

In the US/Imperial equation, the units are hours, feet, degrees Fahrenheit, and British thermal units (BTU) as discussed above.

The metric R-value equation shows degrees Kelvin, area in square meters, divided by energy in watts. (The two energy units are slightly different: Watts are units of energy per hour, where BTUs do not contain any time factor so hours appear separately in the US/Imperial equation.)

Assigning Insulation (R) Values

Find the approximate area and R-value of each of the following surfaces:
- Area 1 (walls)
- Area 2 (windows)
- Area 3 (doors)
- Area 4 (ceiling/roof area)
- Area 5 (floors over a crawl space)
- Area 6 (slab floors and basement walls)

To make your calculations, you'll need to assign R-values to materials already in place. Below, we give estimates for the insulation value for common applications. (The table gives Imperial/BTU-compatible R-values; if you can't find a metric R-value table, you can convert between US/Imperial and SI/metric units of R-value.)

R-value estimates — Walls:
- R5 (board and batten, non-insulated)
- R20 for insulated 4" stud/5.5" thick walls (fiberglass or rock wool batting, drywall, sheathing)
- R30 for insulated 6"stud/7.5" thick walls (as above)
- R36 to R50 for plastered straw-bale (18" to 20") walls

R-value estimates — Windows:
- about R2 for single-pane glass
- R4 or R5 for double-pane
- up to R10 for triple thermal windows with insulating drapes

Be aware, though, that solar gain can offset heat loss from sun-facing windows in many climates.

R-value estimates — Doors:
Common R-values for exterior doors may be:
R15 for an insulated-core, metal-skin door;

R2.2 for a solid wood door 1¾" thick.

(Note that windows on an exterior door are still windows; use the door numbers just for solid areas of the door(s).)

Door figures from www.archtoolbox.com/materials-systems/thermal-moisture-protection/rvalues.html

R-value estimates — Ceiling/Roof:

The engineering toolbox site suggests adding 15% to ceiling/roof heat loss to account for radiant heat loss (the clear night sky is VERY cold, compared with the earth's surface or nearby trees and buildings). Unheated attics may moderate this heat loss to some degree.

- R5 for uninsulated joist and drywall ceiling
- R30 for insulated 6" and R40 for insulated 8" joists (large, foil-backed fiberglass or rock wool batts with the edges properly tacked down to limit air movement)
- up to R50 for double-layered fiberglass batting or 24" of loose cellulose.

R-value estimates — Floors:

Ground temperatures are usually milder than winter air temperatures, so plan accordingly.

Doing a bit of home improvement *before* starting on your heater can have a big effect on your final project. Well-insulated crawl spaces can effectively make the floor an interior wall instead of exposed exterior surface, thus significantly reducing your heating loads.

- R5 to R10 insulation board is commonly used for foundation or skirting insulation.
- R20 or R30 batting can be used between floor joists in suspended floors.
- For slabs, R1 to R2 for a 4" slab, or .1 to .5 per inch. (Definitely use ground temperature when making your calculations.)

Example Calculation

Here are the calculations you would do for an example house that is 36 feet long by 24 feet wide (or 860 square feet):

Area calculations:

R30 Walls:

(36 + 24 + 36 + 24 ft wall lengths) × 8 ft wall height minus (Window area) = (120 ft) × 8 ft − 71sf = 889 sf

R5 Windows (double-pane):

(3 windows 6 ft × 3 ft) + (4 ft × 3 ft) + (2 ft × 2.5 ft) = 71 square feet

R36+ Ceiling/roof:

36 ft × 24 ft = 864 square feet

R10 Floor:

36 ft × 24 ft = 864 square feet

We are almost ready to run our heat-loss calculation, but we need the temperature difference (T) between the inside and outside surfaces. This is affected by climate, of course, but it can also be changed by adjusting indoor temperature, or by buffer zones (unheated rooms that reduce the temperature of the exposed walls).

For a day that averaged 30°F, with the ground temperature a little warmer at 50°F, and we kept our home at 70°F indoors:

[(Area1/R-value1) + (Area2/R-value2)…] × (indoor temp. − outdoor temp.) × time = Heat loss

giving:

[(Walls: 889 ft^2/R30 (R units are 1 hr·ft^2·°F/BTU) + Windows: 71 ft^2/R5 + Ceiling: 864 ft^2/R36) × (70°F−30°F =

40°F + [(Floor 864 ft²/R10 × (70–50 = 20°F)] × 1 hour = Heat loss

[(Walls: 30 BTU/hr °F + Windows: 14 BTU/hr °F + Ceiling: 24 BTU/hr °F) × (40°F)] + [Floor: 86 BTU/hr °F × 20 °F] × 1 hour = 4000 BTU (for one hour), or (multiplied by 24 hours) 96,000 BTU/day

Is that typical for an 860-square-foot house? It depends on your climate.

If the day averaged -10°F (with the ground temperature remaining the same), the difference in temperature would be 80 degrees F instead of just 40:

[(Walls: 30 BTU/hr °F + Windows: 14 BTU/hr °F + Ceiling: 24 BTU/hr °F) × (80°F)] + [Floor: 86 × 20°F] × 1 hour = 7160 BTU/hr, which would be 172000 BTU/day (not quite double the previous example)

In practice, local building standards often include more insulation in colder climates, and it's easy to see why.

Heat Loss by Ventilation

The heat loss equations in this section and those that follow can be found, with nicer notation, on Engineeringtoolbox.com. Search their site for "Heat Loss from Buildings."

Heat loss by ventilation (fresh air delivery into buildings, mechanical or otherwise) and infiltration (expected natural movement through wall cavities, roof vents, cracks, porous materials, etc.) are roughly similar.

The heat loss due to ventilation without heat recovery can be expressed as:

$H_v = c_p \rho q_v (t_i - t_o)$

where

H_v = ventilation heat loss (W)
c_p = specific heat capacity of air (J/kg/K)
ρ = density of air (kg/m³)
q_v = air volume flow (m³/s)
t_i = inside air temperature (°C)
t_o = outside air temperature (°C)

The heat loss due to ventilation with heat recovery can be expressed as:

$H_v = (1 - \beta/100) c_p \rho q_v (t_i - t_o)$

where

β = heat recovery efficiency (%)

Modern heat recovery ventilation systems can achieve efficiencies of 50 to 80%.

Heat Loss by Infiltration

Infiltration includes air movement through leaks or porous materials in the building construction, opening and closing of windows, gradual upward movement of warm air escaping through ceilings and attics, etc.. As a rule of thumb, the number of air shifts is often set at 0.5 per hour. The value is hard to predict and depend of several variables—wind speed, difference between outside and inside temperatures, the quality of the building construction, etc.

The heat loss caused by infiltration can be calculated as:

$H_i = c_p \rho n V (t_i - t_o)$

where

H_i = heat loss infiltration (W)
c_p = specific heat capacity of air (J/kg/K)
ρ = density of air (kg/m³)

n = number of air shifts, how many times the air is replaced in the room per second (1/s) (0.5/hr = 1.4 ×10^{-4} 1/s as a rule of thumb)

V = volume of room (m^3)

t_i = inside air temperature (°C)

t_o = outside air temperature (°C)

Heat Loss by Air Exchange

The next section appears on the website www.sensiblehouse.com. We include it here, not to discourage anyone with scary-looking calculations, but to show how heat designers think about and use these equations. (See website for more detail on modeling home heat loss in general: www.sensiblehouse.org/nrg_heatloss.htm, viewed 5/27/2015.)

Ideally we'd like a table or formula that would allow us to know the value of ACH [air exchanges per hour] at various temperatures so that we could get a more accurate total heat loss for any outdoor temperature, but because wind speed is typically such an important component of ACH, it's really impractical to do this; hence, we use the estimates.

The issue here is that we know that the infiltration rate is higher when it's cold and windy, so unless the typical weather conditions are that it's windier when the temperatures are moderate than cold, ACHnat will result in underestimating the heat loss due to infiltration. If you're calculating the worst case heat loss (for example, for equipment sizing) rather than using ACHnat in the heat loss formula, you might want to use ACH50/10 or even ACH50/5, which are both just different fudge factors than those used to calculate ACHnat — it really depends on your climate and whether you think ACHnat is a good estimate of infiltration at cold temperatures or not.

Once you've determined an infiltration rate, the heat loss is calculated via one of the following simple formulas:

$$Q = V * ACH_n * .018 * \Delta T$$

or

$$Q = CFM_n * 1.08 * \Delta T$$

Where Q is the hourly infiltration heat loss, V is the volume of the house, .018 is the heat capacity of air (whose units are BTU/ft^3-°F), *ΔT is the difference in temperature, and ACH_n and CFM_n are the normalized blower door test values for whatever conditions you want to assume — the typical assumption being ACHnat, i.e., the adjusted value based on the statistical model representing the "natural" ventilation rate. Note that the value 1.08 is just the heat capacity of air .018 times 60, since CFM is a per minute rate and we're looking for a per hour rate. [*Author note: CFM is cubic feet per minute.*]

Intuitively, this number represents the amount of heat contained in the air that leaks out, or more appropriately the amount of heat required to heat up the air that leaks in as a result of air leaking out. There is some evidence that as air leaks out through an insulated cavity, the cavity acts a bit like an HRV (heat recovery ventilator), but given that you don't really want air leaking through an insulated cavity, especially not at a slow rate where condensation can happen, it's best not to assume this happens.

If you have mechanical ventilation, the calculation is essentially the same, but in this case the ventilation rate is whatever the fan is rated for. If the ventilation is a HRV or ERV, you need to adjust the temperature difference by the efficiency of heat recovery; so, for example, if ΔT is 50 degrees and the efficiency is 70%, then the effective temperature difference is only 30% of ΔT, or 15 degrees in this case.

Infiltration rates are going to vary with wind speed, difference between indoor and outdoor temperatures, and how the house is built and detailed. We can't control the wind speed, but we can buffer it with windbreaks or unheated outdoor storage areas.

We can control the indoor temperature, but we don't always want to. Remember, though, that you can lower the heat loss while still enjoying comfortable temperatures by keeping the heat away from outside walls. Barriers such as quilted hangings, storage closets and cabinets, or closed rooms can keep you warmer yet allow less heat to reach exterior walls, meaning less heat gets lost to outside.

We can definitely improve on how most houses are built and detailed. Keeping the house in good repair helps a lot. Wall penetrations for electrical outlets, some types of lighting, doors, windows, and corners are common places to find air leaks. Attic hatches can represent relatively huge heat-loss holes, and can be improved with well-sealed and insulated hatch doors (or completely sealed over, replaced with an outdoor access through the gable wall). Some owners re-sheathe the house (be careful to use breathable materials that are compatible with the structural detailing of older houses; vapor barriers should go on the warm side of the wall). Check behind ceiling and floor moldings for badly sealed gaps to fix.

Sometimes we want to change the temperature of the house using outside air. Cross-ventilation (opening windows on upwind and downwind sides of the house) is a great way to move air through the building when outdoor temperatures are more comfortable than indoors, cooling a hot house at night, or freshening up the rooms on a summer morning before shutting things again for the day. Air movement may also contribute to evaporative cooling, or affect condensation problems (warm moist air deposits water on cold surfaces, but cool air can remove moisture from a warm house).

Moisture content of the air is not considered in the above equations, but it affects the heat capacity of the air, how much heat is lost (or required to re-heat incoming cold, moist air), and most importantly, moisture affects condensation levels which may be a bigger problem in leaky walls than just wasted heat.

Limiting Heat Loss

The most common (and cheapest) ways to reduce heat loss are:

- add insulation at the easiest or least-insulated areas by adding lined drapes or

closed-cell blinds for the windows and another layer of insulation in the attic or ceilings; create better seals around attic hatches.
- lower the average indoor temperature if possible (for example by closing off unused rooms, creating cooler "zones" around the edges).
- add more insulation around the perimeter of slabs or crawl spaces, and against floors or ceilings.
- check for cold air leaks (and possible moisture condensation) around electrical outlets, windows, and other wall penetrations. Walls are harder to insulate without disrupting the household, but it's worth doing when possible.

The heating calculations above offer some big clues to how we can reduce heat loss to the house as a whole. But they are based on a typical American assumption: that we are going to build a big stud-framed house and heat it by filling it with warm air. There are a lot of other approaches to staying warm, some of them radically more efficient.

The next section gives some simple, sensible suggestions — no math required.

Home Design for Heating Efficiency

Present-day American households devote 30% to 75% of their energy to heating and cooling. The first step to reducing those bills is not necessarily changing the furnace, but making the house easy to heat.

Modern North American homes were largely designed during a 400-year surplus of cheap fuel. On top of that, the average home in the US has doubled in size just in the last 50 or 60 years, yet fewer people now live in each home. Old-world cultures that have survived fuel shortages and famines know better. Maintaining a large house can bankrupt the owner. Old money tends to find itself at home in small, elegant rooms and apartments, furnished with beautiful and durable materials, rather than bleeding away as monthly utility bills to heat a lot of glass and plastic.

One of the fundamental insights of rocket mass heater is "heat people, not space."

When my wonderful big oak office desk keeps my feet away from the heat, I don't have to heat up the whole house to get warm. I can go sit on the rocket bench. If I really need to concentrate at my desk, I can go make myself a little lump of heat, like a cup of tea or a hot brick wrapped in a towel, and take it back to my desk. I can take these little warm lumps anywhere: in the car, out into the woods, put one on my sore back like a heating pad.

Cupboard-beds, four-poster beds with top and side curtains, and even blanket forts can be a way to create a personal warm zone in a cooler house. Social radicals might even consider switching bedrooms, so that the hottest one goes to the person who is always cold, rather than assigning them by size or status.

So when we talk about heating our homes, what are we really trying to achieve?

At the core of the house, there is food, water, fire, and sleep.

All of these can be accommodated in a single room if necessary; or in a core of no more than three or four rooms (living, master bedroom, girls' and boys' bedrooms). Kitchens generally heat themselves. Optional "spare" rooms should be optional to heat. If you spend less than 40 minutes per day in the bathroom, consider a separate on-demand heater such as a heat lamp or small radiant wall-heater.

Elements to a more energy-efficient household:

- Scale down. Look at camper and live-aboard boat designs for inspiration. Now take a look at your house — can you condense the daily activities into a smaller heating and cooling footprint, with "spare" rooms closed off until needed? Art Ludwig's suggestion is about 200 square feet per person, maybe 300 square feet if you're still getting used to the idea of scaling down. (Art Ludwig, OasisDesign.net, "Can a 4000 Square Foot Home Be Green?" http://oasisdesign.net/faq/green4000ft2home.htm)
- Share. Consider renting spare rooms, or building duplex-style townhomes rather than isolated houses. Exterior walls contribute to heating load; interior partitions don't. Bunk-beds rather than separate rooms can bring more kids closer to the heat.
- Stack and compact. Apartments or bedrooms on upper floors can pack a lot of living space inside less exposed surface. Stack functions in the same space, using unheated storage for projects and tools until you need them.
- Group and zone. Bring stuff that needs to be warm into the core — people, plumbing, batteries, delicate plants or pets. Close off seldom-used rooms, and heat them only as needed. Organize your storage along the outside — closets, cupboards, and pantries, screen-porches, sheds, or mudrooms, storage racks. Most goods keep better when dry and cool anyway.
- Heat upward. Locate the heat low in the occupied rooms, and use thermal mass to store it. Radiant-heated floors or low benches counteract air stratification (the tendency for hot air to collect at the ceiling). With a radiant floor, people report the same comfort in a house up to 10 degrees cooler, a massive reduction of heating load. Look up k'ang, ondol, hypocaust, and tawakhaneh heaters for time-tested examples of heated floors.
- Don't go too low. Putting radiant heaters in an unused space, such as the basement, can lose more than half your heat into the cellar walls and soil. (A furnace or boiler is designed for the inefficiencies of heating at a distance; radiant woodstoves, fireplaces, or masonry heaters are not.)
- Insulate and weatherize. Just as animals put on a thicker fur coat in the wintertime, we should have a good layer of insulation around our house. Make the easiest and most effective measures your priorities; these might include attics/ceilings, attic hatches, north-facing windows (install heavy drapes or closed-cell blinds), crawl spaces and any exposed floors, or any uninsulated areas that desperately need attention.
- Thermal mass. If insulation is the fur coat, thermal mass is the dense blubber and bone that keeps Arctic wildlife warm. Use stone, brick, jars of water or oil, or any other dense materials to store heat in the core.
- Ventilate. The house and everything in it needs to breathe. Allow for a minimum of one third of the home's volume of air

per hour. Half the volume is more common, and some houses draft one to three times their volume per hour. The minimum ventilation provides fresh air and prevents mildew, mold, and structural rot due to condensation. Provide controls for optional ventilation — windows on opposite walls for a cross breeze, stairwell doors to let air circulate upward or trap heat on certain floors, heat-exchanging ventilators rather than letting air come in at random through dusty crawl spaces and drafty walls.

- Isolate spare rooms. For rooms that are not consistently used (at least 4 hours per day), consider isolating them from the main home with doors and interior wall insulation, insulating them well if you need to be able to warm them up on demand, and using a space heater or small parlor stove when needed. Hang quilts on the walls and use a hot water bottle or warm, wrapped brick to warm up the guest bed. Bathrooms can be heated on demand with a heat lamp, and kept half as warm as the rest of the house (50°F/10°C — not freezing). Spare rooms on the shady or windy sides of the house can waste a lot of heat, or they can become a protective buffer that saves the rest of your house.

Climate-adapted Housing

Historic houses last because they are worth keeping. Usually, they work well. Pre-industrial buildings, especially those still in use, can offer good examples of appropriate design for your climate and culture.

However, don't be misled by parlors and museum tours or by fantasy designs. Many famous houses display only their elegant entertaining rooms that were designed for special occasions; visitors may never see the everyday core of the house (a parlor and a servants' kitchen, for example, might be the only rooms that were routinely heated on ordinary days). Maintaining an *orangerie* has always been a colossal expense, and therefore a showy form of conspicuous consumption. "Unique showpieces" may be an architectural fantasy, but if they remain unique, it could be because they were not comfortable enough to be copied.

Farmhouses often take a more practical and thrifty approach to heating and multi-use space. Migratory or nomadic cultures honed their households down to high-quality essentials, like climate-adapted clothing, fur bedding, durable containers, and easily assembled small tents or sleeping shelters.

Other interesting designs include modern energy-efficient and passive solar designs, and "tiny houses" as described in *Tiny Homes: Simple Shelter* by Lloyd Khan, 2012, or Lester Walker's *Tiny Tiny Houses, or How to Get Away From It All,* 1984 (two favorites from over a hundred recent books showing tiny house plans; it's a big thing lately). Tiny homes are not actually that efficient, they have a lot of surface area to insulate compared to their footprint, and one or two people living alone can almost never equal the efficiencies of a bigger, shared household. But a little jewel-box cabin provides personal creative control, privacy, maybe even a little sanity, and can be a nice retreat from the demands of social life.

It's easy to fall in love with a certain style, and forget that the building exists to perform specific tasks for your daily comfort.

Does the house fit your climate, lifestyle, and expectations (health, wealth, old age)?

Now is a great time to figure out the natural function of all those towers, gables, colonnades, and cupolas. Do they suit the climate? If your house is too big or has too many picture windows to heat affordably, you may want to create a cozy little core in one corner for everyday living, and let the rest of the house fend for itself. Some fancy older houses had extra rooms that were explicitly *intended* for seasonal use only, for special occasions, or for sunny days; I wonder if their designers would laugh to watch us trying to heat every room in every season.

If you love a particular style, consider the climate where it belongs. Russian, Polish, and Swedish immigrants brought some great cold-weather technologies to the North American Midwest. English, French, and Italian influences dominate the coasts. Spanish and Pueblo peoples contributed remarkable architecture to the Southwest. Japanese, Italian, and Peruvian styles share notable symmetry and interlocking structures, which can be valuable in earthquake zones.

"There is no such thing as a perfect house," said Gran'ma Enid McCaffery Ritter, who lived in 32 different homes from the 1930s Depression through the 2008 recession. There's always an awkward corner, stairwell, or missing pantry. Don't let imperfection rob you of enjoying the good points.

In shelter design as in life, success comes from good observation skills, knowing your priorities, practice and improvement, and a good sense of humor.

Our booklet, *Simple Shelter*, has even more ideas for how to stay warm, at all different scales (personal, room, household). The booklet is available at http://www.ErnieAndErica.info/shop.

Chimneys

Some materials about rocket mass heaters give the impression that "you don't need a chimney." Unfortunately, this is not true in most actual homes.

Hot air rises; cool air falls. This process ultimately drives almost all the winds and air currents on Earth, from foehn winds to wildfires to a candle flame. But we don't really think much about it; and even when the emergency woodstove in the basement belches smoke in our face on the same miserable evening that an ice storm knock out the electrical lines, we don't necessarily make the connection.

Some chimney facts:

- Cold chimneys do not draw (or, they draw downward).
- Horizontal "chimneys" do not draw in either direction.
- Taller, hotter, smoother, more vertical, more airtight chimneys will draw like a jet.
- Cooler, shorter, rough, convoluted, zigzag, or broken chimneys will draw unreliably, if at all.

What do you think happens if you connect a bundle of chimney-like shapes, like a stack of warm rooms next to a stack of vertical 2x4 studs, some sewer and appliance and roof vents, a couple of woodstove chimneys, and let them all draw room air from the same house?

You will get a complicated mess! It's like trying to pour tea from a pot with three spouts. The losers (the lower, cooler holes) serve as inlets for the taller, warmer openings.

A chimney works by facilitating upward, convective flow of warm air or smoke, while preventing turbulent interaction with surrounding air. (If the chimney is too large, or rough, then turbulent flow can take place inside the chimney and severely hamper the draft.) It provides a barrier between the cold air outside and the hot gases rising inside the stack.

A house also is warmer than its surroundings, and provides a barrier between its internal warm air and external cold air. If your house did not allow at least a small amount of air movement, you would suffocate. During the winter, your house is naturally a pretty good chimney.

In warm buildings, as in chimneys, the warmest air rises upward. At the top, it tends to leak outward (positive pressure). Cooler air comes in at the bottom (negative pressure). Somewhere around one third and one half the height of the wall, in theory, there's a neutral zone where air neither comes in nor goes out.

This means that any opening in the lower third of a home is almost guaranteed to act as an air inlet, all other things being equal.

Exhaust fans for range hoods, bathrooms, ventilation systems, and appliances can create even more negative pressure inside the house, increasing the zone of negative pressure to the point where ANY hole in the building may draw inward instead of out. Airtight homes are worse for generating negative pressure than older, leaky homes, because lots of appliances vent outward, often without replacement air.

Air doesn't read any kind of directional arrows on the chimney; it just flows from areas of greater pressure to areas with lower pressure. So if your house has negative pressure inside, there's a good chance that air or smoke will flow inward through any available hole, including the one you thought was an "exhaust chimney."

It's our task to ensure that both exhaust fans and chimneys are able to function properly. This requires some vigilance:

- Mitigate any chronic negative pressure problems in the house, for example by providing warmed air to replace air being drawn out.
- Avoid operating at cross-purposes. For example, if someone upstairs is too hot and opens the window, and someone downstairs is too cold and stokes the stove, you can burn for hours without fixing either problem. Train savvy operators in the use of simple controls (like doors).
- Don't burn unattended. A good operator is ready to adjust make-up air, close room doors, or even open an upwind window if a bunch of housemates all start running fans at once. (See Appendix 3 for more about make-up air.)
- Be leery of basement installations; the lower in the house you go, the greater the negative pressure can be. Cold chimneys to basement emergency stoves or fireplaces are notoriously likely to produce "smoky smells" due to downdrafts.

Keeping most of the chimney inside the house gives the reassurance that, with the exception of a few feet at the top, the chimney won't be colder than the house when you try to start the stove.

If the exhaust itself is considerably warmer than outside air, an insulated chimney can maintain that difference well enough to draft upward. But insulation does nothing if outside air is warmer than the exhaust. "Priming" or preheating a chimney can get the draft working better.

In our rocket mass heaters, the firebox itself makes a significant contribution to draft. The combustion unit is a thermosiphon draft pump, with somewhat self-regulating draft; a hotter fire draws well, yet the stove doesn't draft as much once the fire goes out.

In a tiny house with plenty of incoming air and no competing draft forces (wind, roof vents, etc.), this thermosiphon effect is strong enough to push exhaust horizontally at least 20 feet. This is remarkable to anyone who is accustomed to wood heat. Most woodstoves require a minimum chimney length of about 8 vertical feet. Most other wood-fired heaters (including masonry heaters) rely on the warm exit chimney to power the draft of the system; this is why most masonry heaters are shaped like a column or vertical wall. Rocket mass heaters effectively have their 8-foot chimney folded in half, and located right at the firebox … but is it enough?

In larger or taller houses, and in windy climates, we run into problems. The simple fact is that for a conventional home, the easiest and cheapest *reliable* solution is a conventional chimney.

Well-designed chimneys and homes can reduce building pressure problems and wind gust problems. The principles for good chimney design are pretty well described in the building codes. They must be two feet above anything within ten feet (to avoid skirling winds around nearby buildings) and three feet above the roof (to avoid swooping currents riding up and down the roof).

In areas with heavy snow loads, the required chimney height may be taller — as much as 5 or 6 feet above the ridge in some areas. (Do your winter storms ever deliver 4 feet of snow at once?)

Another chimney design issue is how to get the chimney safely out of the house without causing leaks or fire hazards. The first choice is location: through the roof, or through a wall?

Through-roof chimneys near the ridge are our favorite choice. The building protects and supports the chimney, you don't need weird brackets or lots of expensive insulated sections, and the top of the roof is the easiest place to flash the chimney so it's leak-proof and stays that way. (Avoids ice dams, excessive rain runoff, and other problems found nearer the eaves and dripline.)

However, in many houses the ridge location requires passing upward through ceilings, floors, and rooms. If the doorways line up wrong, there might not be a suitable place available without some reframing. Midway or lower down the roof can work OK if your climate doesn't get much snow, or if you can build a blind dormer to divert snow and ice.

Through-wall chimneys often seem easier to novice builders. They are easier to reach, and the through-wall parts themselves (two collars and an insulated section) are often cheaper than a through-roof kit. But the through-wall opening in itself usually isn't enough. Chimneys should never open directly below the eaves in any climate: you are unlikely to get good draft due to intense pressure gusts, and it could start a fire. So we have to get up above the eaves.

In order to bring the chimney from that side exit up above the ridge line, there will be a lot more exposed chimney than if we'd gone through the roof. To avoid cold-start problems, water condensation, and creosote deposition, the exposed chimney must be well insulated.

Insulated chimney sections are more expensive and heavier than the single or double-walled sections (which can be used indoors, as long as there's enough clearance, since indoor stovepipe stays warmer than outdoors). So now, not only are we buying a lot more of the expensive, heavy-duty stuff, but we need to somehow support it going up a long ways outside the house. And if you read the manufacturer specs, we really should be building an insulated chase around that expensive insulated chimney if we want it to stay warm enough to perform.

Basically, putting up a chimney outside the house means building it a house of its own so it can stay warm and not have to cling to the outside of the house like a monkey.

Even with all that, the chimney still won't draft as well as it could have if it was sharing its heat back and forth inside your warm house.

Through-wall installations can also have problems with water running down the chimney and back into the wall. Storm collars help a bit, but it usually takes several. Coming out as high as possible on the gable wall, rather than under a drip line, helps with several of these issues, but it is still not as easy or weatherproof as a through-roof installation.

On the whole, it's best to install a full-height chimney, through the roof, near the center of the house.

Because many types of fire hazards build up over long time periods (such as long-term baking of woodwork that increases its flammability, or the corrosive effects of smoke), building codes are a more reliable guide to best practice than an individual's personal experience. Very experienced builders often build in *excess* of code requirements, rather than to the bare minimum.

Exceptions:

The original rocket mass heaters were built in cob (earthen-walled) cottages, in a valley with mild winters and consistent up-and-down-valley winds. These buildings also had no roof vents, being covered with a pond-liner and living roof (with 100% cedar mill-waste ceilings and cedar purlins, in case that sounds like a recipe for rot).

In this situation, the walls are noncombustible, there are no snow loading issues, the house acts more like an oven than a chimney, so the builders have gotten away with little snorkel-like chimneys poking up just over the eaves. If your house is one story, has a ventless membrane-lined roof, and you live in a place with consistent wind directions so you can put your exhaust on the downwind side, you might get away with it.

Further discussion of chimneys, and commonly proposed alternatives like fans or outside air, can be found in Chapter 2 and Chapter 6.

Appendix 5:
Wood Heat Considerations

Heating with natural fuels is an ancient part of human life. Recent industrial centuries have turned fire into an automated specialty, allowed a large population to grow up without personal experience of wood heat (or other natural fuels, such as peat, dung, or straw).

Our 2013 publication *The Art of Fire* contains a great deal of back-to-basics information about managing fire in general. It is available at www.ErnieAndErica.info/shop, and www.rocketstoves.com.

For serious home heating, several additional aspects of wood heat bear mention.

The Nature of Fire: Fire is a chemical reaction that, once started, is self-perpetuating — so long as fuel and air remain available. It has its own responses to changing conditions, such as wind, fuel quality, and surroundings. Fire does not respond well to wishful thinking; in order to use it effectively, you must understand it on its own terms.

Minding the Fire: Many ancient cultures had a "hearth goddess" who tended the home fires. Watching, listening, and arranging the hearth is a sacred responsibility; it calls for mindfulness — a constant awareness of one's surroundings even while attending to other chores or company.

If heating exclusively with wood is your goal, then you must come to terms with spending a certain amount of time with the fire. Responsive radiant heaters (such as most iron woodstoves, radiant fireplaces, and blower-style heaters) require a live fire to produce heat. Storage heaters, on the other hand, retain warmth long after the fire goes out, but require a certain amount of tending during and just after the fire.

Fuel Handling and Storage: There is an old saying about firewood — that it warms you twice: once in the cutting and once in the burning. Gathering enough firewood is a significant chore. Anyone contemplating a wood-fired heater as a way to save money should definitely include the cost of delivered firewood as an annual expense. A good woodshed is essential; it needs to be large enough to store at least two years' supply of fuel, and ideally have room for tool storage and kindling splitting as well.

Dry Fuel: Fuel must be dry to burn properly. Wet fuel releases large amounts of steam. The process of boiling wet wood costs energy, plus the volume of steam involved (1700 times the original volume of liquid moisture) tends to extinguish the secondary combustion process, depriving the fire of over half its heat. Wet wood causes creosote deposition, which can lead to chimney fires. Depending on how wet a fuel is, it may be difficult or impossible to light a fire. Properly dried and seasoned wood has 15–25% moisture content, most of which is "bound" moisture in the form of volatile resins. (Source: www.chimneysweeponline.com)

Fire Control: Ancient civilizations had a very healthy fear of fire and a tradition of vigilance (sometimes extending to rooftop platforms equipped with tanks, for wetting down urban roofs in time of fire). Fire that escapes its hearth can quickly become a deadly force. Modern building code prescribes clearances, spark arrestors, building practices, and other measures to ensure that a fire remains contained within its designated fireplace and does not spread to other parts of the building.

Smoke Mitigation: Smoke is a secondary danger of fire, one that is toxic and potentially lethal. Although smoke inhalation kills many victims of a house fire before they are ever touched by flame, only in recent centuries have we expected to keep indoor and outdoor air smoke-free. Because smoke is unburned fuel, any fire that produces large amounts of smoke is inherently inefficient. Clean fire is much healthier, and offers a better return on investments in fuel and tending.

Hearth Inspection and Maintenance: Anyone depending on wood heat needs a reliable hearth. Masonry and metal components must remain in good condition, with several layers of protection between the fire itself and any surrounding combustible materials. Hearth inspection should be routine before starting any fire (to catch misplaced combustibles on stovetop or mantle). Chimney flue inspection and cleaning should occur at least annually. Homeowners new to wood heat are well advised to hire a professional chimney sweep for the first inspection, and get a personal lesson in what to look for with their particular chimney.

Hidden and Long-term Fire Hazards: Good design for safe combustion requires a knowledge of materials and building history beyond what is normally expected of an individual person.

In most circumstances wood doesn't catch fire below about 400°F (and anyone who's tried to light a campfire knows that logs don't catch easily without a lot of help). Yet wood that is routinely exposed to temperatures above 150°F can undergo a long-term baking process known as *pyrolysis,* which breaks down and darkens the wood fibers, creating a very dry and flammable material that can ignite or store embers much more readily than new green wood.

Metal stovepipe reliably carries smoke and heat out of the house — until it is coated with creosote and catches fire. Modern chimneys are built to withstand at least one chimney fire; those who burn wet wood and endure repeated chimney fires may push these materials beyond the limits of their design. Even the most fireproof materials can warp or leak if abused long enough.

Builders who rely only on personal judgment and experience sometimes skimp the clearances to combustibles, reasoning that their day-to-day experience defies a fire to catch across so much material. Those who've poked deeper into historic buildings and city fires take fewer chances.

Fire does not make exceptions for appearances. An exposed chimney wall is in many ways safer than an enclosed or walled-in chimney, since the warm void spaces hide and nurture problems. We make it a policy never to disguise one material as another, for example painting over lower-grade metal to make it look like enameled stovepipe. Fire safety is nothing to fool around with.

Flame Dynamics

Temperatures

Clean fire starts around 1100°F, when carbon monoxide can be completely oxidized into carbon dioxide. For our purposes, we aim for a firebox temperature between 1200 and 2400°F. Temperatures above 2600°F not only require rarer and more expensive refractory materials, but can also increase pollution by burning nitrogen from the air.

The process of fire happens in stages. First, any remaining moisture is driven from the wood as steam. (This can have the effect of suppressing the fire if there is enough of it.) Second, volatile chemicals in the fuel vaporize and break down, creating a highly flammable vapor that burns as flame. From 50–70% of the fire's heat comes from these volatile oils, resins, and vaporized carbohydrates.

Third, the remaining solid material of the fuel "catches" and begins to burn as coals or embers. This third stage is the one that makes fantastic barbecue, because the heat is released as a steady glow rather than dancing flames. Embers generally produce less smoke, though invisible carbon monoxide may be produced if the temperatures become too low.

The goal in creating clean fire is to ensure that all the volatile combustion gases remain hot enough, with ample air, and ample mixing, for long enough to burn completely. Ianto Evans refers to this in his book, *Rocket Mass Heaters,* as "time, temperature, and turbulence."

Chemistry

Clean fire produces two main chemicals: water vapor and carbon dioxide. While this "cleaner" exhaust is oxygen depleted and not safe to breathe in an enclosed space, it's much less toxic than carbon monoxide or smoke.

Smoke is unburned fuel. A smoky fire is by definition wasting fuel (unless you are trying to smoke meat or signal for help). The unburned or partially burned fuel in smoke can include tars, oils and resins, soot particles, kerosene and turpentine, wood alcohols, vinegar, carbon monoxide, and many other chemicals. Most of the chemicals in wood smoke have two things in common: they are flammable, and they are not good to breathe.

Carbon and Creosote: Why Hot Fires Are Safer

Clean Fire = Clean Chimneys and Clean Air

If smoke passes through a cool chimney or heat-exchange channel, it will condense as soot (carbon) and creosote. Creosote is a tarry residue that is difficult to remove from chimneys, highly flammable, and burns very hot. Creosote fires can demolish steel chimneys, cause masonry to crack, and char or ignite adjacent woodwork. To reduce creosote buildup, metal woodstoves are required to keep their exhaust-chimney temperatures above 350°F (180°C). This encourages the smoke to leave the chimney without condensing on the sides (a cold chimney acts as a creosote distiller if any creosote is present in the smoke).

This hot minimum exhaust temperature represents a lot of lost heat. A masonry heater or rocket mass heater could theoretically be 20–40% more efficient that a woodstove, just by reducing the exhaust temperature. Lower exhaust temperatures mean less heat is wasted outdoors, and more heat is being saved inside the building. (In the case of masonry heaters, their ability to store heat long after the fire goes out with the exhaust completely shut down is an efficiency factor that is rarely considered in the conventional comparisons. Masonry heaters give off no emissions and virtually no wasted heat during the stored-heat phase.) However, to reduce exhaust temperatures safely, the heater must burn virtually smokeless, eliminating creosote before it can condense inside the pipes or chimney.

Woodstove efficiency ratings take this disadvantage into account; advertised efficiencies are based on a theoretical maximum efficiency of 85%, with 15% of the theoretical heat being released out the chimney to meet regulations. If a stove was able to deliver all 85% of its available heat, while maintaining the required chimney temperatures, it would be considered 100% efficient.

So a "75% efficient woodstove," graded on a curve where 85% = 100%, puts about 64% of the available wood energy into heating the home. It releases the rest as smoke or chimney heat. Burning damp wood, or operating the stove with improperly adjusted air feeds, can reduce the efficiency by half or more of this already-underwhelming total. In ancient times, many poor people's homes used fires without chimneys to conserve more heat in the home; to survive the winter on scarce or costly fuel was a bigger safety consideration than the immense amount of smoke inhalation involved. Clever peasants used every trick in the book to minimize smoky waste and extract the most heat out of the fire.

Many owners don't like losing heat out the chimney. But we also don't want to huddle in a smoky cave, or suffer repeated creosote fires. To safely lower chimney temperatures, we need a consistently smokeless fire.

People have been designing smokeless stoves for thousands of years. Archeological examples of "fox stoves" from the Indian subcontinent appear as early as 5000 years ago. Our ancestors learned this skill as an essential element of survival. But this skill has become uncommon through disuse and cultural displacement in the modern day. Rebuilding it takes intelligent research and willing practice.

The first step to clean fire is dry fuel. Dry, seasoned fuel has about 15% moisture content. To estimate the moisture content of your wood, measure its weight, then bake it in an oven for 12 to 24 hours (or store it indoors for a few months). Weigh it again. The difference in weight represents water driven off. A "lazy" person who fails to build a proper woodshed may be working more than twice as hard as necessary: hauling all that water weight into the house, and then hauling in extra wood in order to provide the energy to boil away the water in the original wood.

If your fuel has not checked (small cracks that appear as wood dries) or lost weight since you stored it; if you see sap boiling at the ends of your fuel in the woodstove; or if there are green leaves, green moss, or fresh mushroom growth in the woodpile, your fuel is not dry. Fuel that rests on the ground, or under a dew-collecting tarp, or is exposed to periodic rainfall is not dry. When you ask a seller if the wood is dry, and they reply, "It hasn't rained in at least a week," the fuel is not dry.

A good woodshed lifts the fuel off the ground, has an overhang to prevent rain and snow falling on the wood, and has ventilated sides to encourage air movement. Green fuel can be stacked loose to dry faster; once dry, fuel wood can be stacked tight.

In a camping or emergency situation, the drier wood outdoors will be "standing dead" fuel (low, dead branches still attached to a living tree and sheltered under the canopy). Dead-and-down fuel lying on the ground will contain much more water. Green branches are so full of moisture, even when the tree is dormant, that they are used deliberately for smoke signals, and for evaporative water collection in desert environments. If the branch bends as you attempt to pluck it, it is not dry. Good kindling jumps when it is split, rings when it is tapped; excellent kindling makes you somewhat ashamed you are not using it to build a violin.

Only the operator can procure dry wood. The stove's designer and builder can do nothing except hope that their work will be appreciated and properly used by an informed and responsible owner.

The second step to clean fire is focusing the heat, fuel, and air together to produce a clean burn. Keep in mind Ianto Evans's "time, temperature and turbulence" rule — you must keep the ingredients hot and well-mixed for complete chemical combustion. An experienced campfire cook can achieve a nearly-smokeless fire using the wood alone to guide and focus the flames. But most people who are burning wood for heat want to spend less time tending the fire.

The heater mason's job is to make it as easy as possible to run a smokeless fire, nearly all the time. The owner's job is to follow the basic instructions, using dry fuel and burning the appropriate size fire for the heater. (When less heat is required, instead of using a smaller fire, the owner can burn the heater less often — but always at proper temperatures for clean fire.)

The best-designed masonry heaters do not require cleaning very often (if ever), because the wood is burned so thoroughly that no charcoal, and very little mineral ash, is left behind. When paper is burned, fly ash will accumulate faster. A typical woodstove is designed so that all the ash either lands back in the firebox or flushes out the chimney. A masonry heater may have an ash grate or cleanout drawer. A rocket mass heater, and some masonry heaters, can be cleaned by scooping the ash out through the same opening where the wood goes in. Annual or semi-annual removal of fly ash from the heat-exchange channels is done with a vacuum or brush and bucket, when the stove is cool.

Air-to-fuel Ratios

Airtight stoves attempt to prolong a fire by starving it of oxygen. While this method (and related techniques like "banking" a fire) can be effective for short-term overnight heating, in the long term, the practice of starving a fire is unsafe and not very efficient.

Air-starved fires that include fresh volatile wood produce large amounts of smoke and creosote. Since smoke is unburned fuel, more than half the heat potential of the wood is lost, requiring much larger fuel storage to heat the same building compared to a more efficient method. And finally, since creosote and chimney fires are very dangerous, stoves using this method require a substantial investment in chimney insulation and creosote cleaning. An airtight stove is basically a tar and creosote retort disguised as a parlor heater.

EPA-certified woodstoves are no longer allowed to be truly "airtight." Most provide a supply of heated secondary air for clean combustion that cannot be reduced by the manufacturer's dampers or air controls. Many still provide a long-burn option where the fire can burn untended for 6 to 8 hours or more.

Fireplaces that run with full-open air (35:1 air-to-fuel-ratios) are exempt from some of the air-quality regulations that govern controlled stoves, on the assumption that they will never be starved.

Rocket mass heaters do not need to starve the fire of air to produce steady warmth. Instead of trying to bring the combustion process down to a comfortable temperature, masonry heaters let the fire burn hot and clean, and store the heat in a large masonry mass for gradual and comfortable release.

Rocket mass heaters in particular run well with open air and a full fuel box: the tight-packed fuel allows just about the right amount of air to enter, for good pre-ignition mixing of hot air and fuel gases. A hotter fire draws more air, self-regulating the balance of air and available fuel. Instead of trying to starve the fire, the operator simply lets it run its course, then closes things down afterwards.

All masonry heaters are typically closed down after the fire is out, to reduce air flow and retain more heat overnight. Some designs, including rocket mass heaters and other "contraflow" stoves, have a peculiar "U-trap" function that slows the flow when not in use. The up-and-down channels act as a thermosiphon pump when the fire is hot, but flow more slowly during start-up and cool-down, and slower yet when there is no fire. The U-trap function doesn't trap 100% of the warm air — that would make the heater difficult to light and to operate safely — but it does help the heater self-regulate and waste less heat when the operator is distracted.

In general, the masonry heater lets the operator off easy, while getting the most from the fire.

Fluid Dynamics
Convection

Convection is the process by which heated fluids rise. It has been discussed at length elsewhere as pertains to chimney function, house draft and negative pressure, and warm air stagnation in the room. For the fire itself, convection is the force which pushes hot exhaust out of the way, and brings cool fresh air in to feed the flames. The right draft speed and amount of incoming air is a critical factor in achieving efficient, complete combustion.

Laminar Flow

Laminar flow is the process by which moving fluids sort themselves into layers of different speeds. The layers nearest a surface are slowed by friction and turbulence. The next layers inward move more easily and faster, as they are lubricated by the air cushion at the outer surface. The central layers in any current move the fastest.

Laminar flow matters in chimneys because the friction component is much greater in small tubes, or rough ones, than in large smooth tubes. Corrugated tubing, or tubes of less than 4″ diameter, can have so much surface friction that only very short, vertical chimneys can draft without complete obstruction of the flow. At a certain lower size limit, somewhere near 4″ diameter, a rocket mass heater as described here ceases to function as described.

Turbulence

Turbulence is when a fluid is mixing and tumbling, not flowing smoothly. Turbulence involves a lot of extra drag and friction, and makes things flow a lot slower. In the fire itself, turbulence can be a positive force, creating good mixing between air and fuel. In chimneys and heat-exchange channels, turbulence is not helpful, and can obstruct draft.

Because abrupt changes in direction create a lot of turbulence, greater volume is needed to prevent flow restriction. Peter van den Berg uses an estimate of 150% of the system CSA as a minimum to avoid flow restriction in areas meeting at right angles, for example the gap needed in the manifold where the gases change direction to enter the bench duct.

Thermosiphons and Chimneys

A thermosiphon is to a chimney what a water siphon is to a drainpipe.

A regular siphon works because of the weight of fluid in the long leg. In the absence of leaks, a lot of water flowing downward can pull a smaller weight of water upward for continuous flow. The siphon works only when there is a significant difference in height between the upper and lower water sources, and the hose remains full of water (or any fluid of about the same weight). A typical siphon is an upside down J, with the short end in the source tank and the long end draining water downward.

A thermosiphon works along very similar principles, except that the general direction of flow is affected by *temperature.* The whole void must be filled with a similar-density fluid, such as air, water, or oil. The hotter fluids will move upward, the cooler ones downward. In a tank or an O-shaped pipe, you can make a thermosiphon by heating one side and allowing the other to cool, establishing a circular flow. A typical room with a hot woodstove is a good example: hot air rises upward near the stove, flows along the ceiling, and cool air flows downward along the opposite wall, especially if it has poorly insulated doors or windows. As a result, the people in the room experience cold drafts at floor level and the stove ends up producing a lot of inaccessible hot air.

A J-shaped thermosiphon (such as our firebox) operates in a similar way as the reverse-J water siphon. The short leg is near the source of fuel and air. The flames follow the longer leg upward, creating very hot temperatures and strong draft. This draws air downward in the short leg. This is why the J-shaped firebox stays cool at the top and doesn't put out smoke: the heat riser keeps it drafting in the right direction. As exhaust gases cool, they can flow downward again.

In addition to a difference in height and the absence of leaks, a thermosiphon also requires a consistent temperature or difference in temperatures. If the short leg became much hotter than the long one, the thermosiphon could stop working or even flow backward. Cold air could flow downward in the longer leg under gravity, as in a cold chimney, or when inside the house is colder than outside air.

If there are leaks, a thermosiphon behaves much like an ordinary siphon, dumping the fluid out both legs at once. In the case of a rocket mass heater, air leaks into the bottom of the firebox will inevitably cause both feed and heat riser to act as two parallel chimneys, causing unwanted smoke and flames to escape out of the feed.

Chimney Draft by the Numbers

For simple vertical chimneys, here are some general equations:

General equation for calculating draft pressure in a chimney:

$$\Delta P = C \times a \times H \times (1 \div T_o - 1 \div T_i)$$

where:

ΔP = available pressure difference, in Pa
C = 0.0342
a = atmospheric pressure, in Pa
H = height of the chimney, in m
T_o = absolute outside air temperature, in K (may also be called T_e)
T_i = absolute average temperature of the flue gas, in K

How much air flows into the bottom of the chimney is estimated by the formula below.

For calculating flow rates (laminar flow, not turbulent):

$$Q = C \times A \times \sqrt{(2 \times g \times H \times [(T_i - T_e) \div T_e])}$$

where:

Q = flue gas flow rate, m³/s
A = cross-sectional area of chimney, m²
C = discharge coefficient (a derived constant, usually from 0.65 to 0.70)
g = gravitational acceleration at sea level, 9.807 m/s² (32 feet per second per second)
H = height of chimney, in meters
T_i = absolute average temperature of the flue gas in the stack, in degrees K
T_e = absolute outside air temperature or starting temperature, in degrees K (sometimes called T_o)

Simple Flow rate:

$$Q = A \times V$$

where:

A = cross-sectional area of flow
V = velocity of flow

Rocket Stove Math Collection

These dimensions are built into the plans in Chapter 4 if you can use standard materials. But for those with nonstandard materials, you may have to adjust things a bit to retain the right cross-sectional flow areas (CSA) and to check that any proposed changes are

A rocket mass heater usually has at least four vertically oriented draft "chimneys" working in different ways:

- the heat riser (major vertical)
- the exit chimney (minor vertical — but the easiest to predict with conventional estimating methods)
- the downdraft column around the inside of the barrel (important downdraft)
- the downdraft in the feed tube itself (negative vertical) — this downfeed chimney is being forced to work backwards against its natural draft; its natural draw is a load (or negative contribution) that the heat riser and other draft elements must carry to maintain correct draft flow.

All of the above chimneys could potentially be considered together as a "system," but modeling their combined behavior accurately through all phases of the fire, and the full range of outdoor temperatures and wind pressures, is almost as complicated as predicting the weather itself. We have put most of our efforts into practical testing in real-life conditions, rather than theoretical modeling. Francesco Truovo of the University of Milan did some interesting work in his master's thesis in 2015, modeling J-type firebox flow patterns using software designed to model house fires.

(Sources: "Chimney Draft" by Jim Buckley of Buckley Rumford Fireplaces, at www.rumford.com; http://www.rumford.com/draft.html. Page dated 5/29/11. Last accessed 6/2/2015; F. Truovo, University of Milan, private conversation, June 2015; and engineering toolbox.com, accessed 6/2/2015)

still comparable to other successful projects. Most builders work from past performance rather than theoretical predictions.

Areas and Volumes

Circles and Cylinders

(R = Radius, D = diameter, CSA = Cross-sectional area, L = Length, W = Width, A = Area)

Diameter: 2*R (usually refers to inside diameter, or ID)

Cross-sectional area (CSA):

$pi \times R^2 = 3.1416 \times R \times R$

Surface area:

Circumference $2 \times pi \times R = pi \times D = 3.1416 \times$ Diameter

Volume of a cylinder or pipe:

$CSA \times L = (pi \times R^2) \times L$

Volume of the barrel outside the heat riser: Area of barrel minus heat riser's area (W × L, or pi × R^2), times height:

$A = (pi \times R_1^2 - pi \times R_2^2) \times H$ or $A = (pi \times R_1^2 - W \times L) \times H$

Area of the heat riser gap — a cylinder wall like a thin paper crown that blocks the space between heat riser and barrel:

$pi \times R^2 \times H$ (H=height of gap).

Rectangles

(L=Length, W=Width, H=Height)

Cross-sectional area: L × W, or L × H

Volume: L × W × H, minus any intrusions

Surface areas: Calculate the area of each side, and add them together.

For a rectangular bench with all sides exposed: 2 × L × W (top and bottom) + 2

× L × H (long sides) + 2 × W × H (short sides). For a bench tucked into a corner, the hidden sides will not require plaster or emit useful heat: L × W (top) + L × H (visible long side) + W × H (visible short side).

Adapting Burn Tunnel Size to a Particular Local Brick

Firebrick stacked in various ways can make burn tunnel courses of certain dimensions easily, without cutting brick. With standard firebrick (2.5" × 4.5" × 9"), we can make courses of these dimensions:.

> 4.5" tall (on edge)
> 5" tall (two bricks laid flat)
> 7" tall (one flat, one on edge)
> 7.5" tall (3 courses flat)
> 9" tall (2 courses on edge)

With half-firebrick (1.25" by 4.5" by 9") available as well as full brick, we can also make these dimensions without cutting bricks:

> 5.75" tall (half-brick flat + 1 full on edge)
> 6.25" tall (half-brick flat + 2 full flat)
> 8.75" tall (half-brick flat + 3 full flat)

We typically don't use gap-filling mortars with good-quality firebricks, but if we wanted to make further adjustments we could include ⅛" of mortar between each course.

Other sizes of brick may stack to slightly different dimensions. You will either need to adjust the dimensions of the firebox slightly, or cut some of the bricks, or make up the differences with a good fireclay mortar.

When adjusting the dimensions of the firebox, roughly square is a good rule of thumb. If your courses will not come out square, consider making your burn tunnel rectangle slightly taller and narrower, rather than wide and low, as the taller shape is easier to clean.

To get the flow area, you multiply width times height for the horizontal tunnel, or width times opening length for the vertical legs (fuel feed, heat riser). Since both height and length are multiplied by the same width, the openings at each end would be the same length as the burn tunnel height.

To calculate the total length of the firebox, I would lay out the bridge (say, four firebricks, about 10" without mortar) and add the opening length on each end.

I check that this length allows the first brick in the bridge, and thus all four bricks around the fuel feed, to stand free of my barrel. With a very large barrel, I might add up to two more bricks in the bridge, and make the firebox up to 25" long.

The heat riser will continue up at least three times the feed height, twice the burn tunnel length.

Critical Gap Sizes: To allow for fly ash, the manifold and barrel should have at least a 1.5 to 2" gap from the outside of the heat riser to the inside of the barrel.

Heat Riser Gap: The gap between heat riser and the INSIDE surface of the barrel top must be at least one quarter the system diameter: 2" for an 8" system; 1.5" or 1.75" for a 6" system. If the heat riser top is broad and not sloped, expect fly ash to clog this area. Ernie prefers this gap be close to the minimum dimensions above. Other builders often make very large gaps.

Heat-exchange Manifold Gap: For the opening where the pipe comes from the manifold into the heat-exchange area, we assume there will be more fly ash and that the gas flow will be mostly from above. So there should be a gap in front of the heat-exchange ducting roughly double the gap for above the heat riser, or the same as the pipe's radius (4" for an 8" system, 3" for a 6" system).

In many manifold designs, there can be a narrow, easily-clogged "slot" where the barrel feeds to the outflow area. It may be easier to envision this curved slot as a bent rectangle. To match an 8" diameter system size, 50 square inches, you'd need an opening roughly: 5" × 10"; 4" × 13"; 3" × 17"; or 2" × 25" (almost half the circumference of the barrel!).

An ash pit in the manifold (extra volume below the entrance to the heat-exchange ducts) reduces the frequency of ash blockages in this critical area.

Why a 4" System Doesn't Work Well: Math-based Discussion

One of the characteristics of laminar flow along surfaces is that there is frictional drag at the surface. Layers farther from the surface flow faster and more freely. In a pipe or channel, the fastest-flowing currents are in the center. The slower side currents provide a sort of lubricating effect, reducing frictional drag on the fast central current.

The whole trick of running exhaust gases horizontally is one of balancing the system's natural draft (which is powered by the fire's heat), to overcome the drag (friction or resistance) caused by moving the gases past the walls of the heat-exchange paths.

To imagine the difference in flow speeds and power that are available with a larger pipe size, it's worth considering the ratios of surface area (which absorbs heat and causes frictional drag) compared to volume (which represents the system's power, in terms of a quantity of fuel and the heated gases that can fit in it and flow through it). The flow rates affect the speed of the system, which in turn affects the fire's temperature and the quantity of heat produced per time.

Surface area (SA) and friction are proportional to the circumference of pipes: $= pi \times D$ $(2piR)$;

Flow is proportional to cross-sectional area (CSA) of pipes: $pi \times R^2$ $(= pi \times R \times R)$:

2" diameter pipe: 2:1 SA = $2 \times pi$ = 6.3 inches; CSA = $1 \times pi$ = 3.1 square inches

4" diameter pipe: 1:1 SA = $4 \times pi$ = 13 inches; CSA= $4 \times pi$ = 13 square inches

6" diameter pipe: 2:3 SA = $6 \times pi$ = 18.8 inches, CSA = $9 \times pi$ = 28.3 square inches

8" diameter pipe: 1:2 SA = $8 \times pi$ = 25.1 inches, CSA = $16 \times pi$ = 50.3 square inches

10" diameter pipe: 2:5 SA = $10 \times pi$ = 31.4 inches, CSA = $25 \times pi$ = 78.5 square inches

12" diameter pipe: 1:3 SA = $12 \times pi$ = 37.7 inches, CSA = $36 \times pi$ = 113 square inches

You can see that for a 2" diameter pipe, the surface area and resulting frictional drag from the sides of the pipe is much larger than the flow area.

Our most common sizes — 6″ and 8″ — have a flow area that is a bit larger than the friction-causing side surfaces.

A 12″ system is more than double the flow of an 8″ system, as a 6″ system is more than double the flow of a 4″ system.

This may also be useful in creating a fudge-factor if you want to split the exhaust into multiple parallel runs, e.g. for a radiant floor, instead of a single linear series of connected channels.

We have found that an 8″ system works OK if split down, for example, into two 6″ straight runs; this increases the CSA from 50 square inches to almost 60 square inches, while the surface area (drag) increases from 25 square inches to almost 40 square inches. The one place that it may be possible to use 4″ pipe is as a lot of parallel, short, straight runs off a larger system. We know of one rather odd system where a 10″ industrial-scale combustion unit (designed to smelt aluminum) was split down into seven 4″ runs to radiate under an earthen floor. The heater runs fine; the CSA is 91 square inches compared to 78.5 square inches, and the powerful fire is apparently enough to overcome the nearly tripled frictional drag (from 31.4 square inches to 91 square inches). The owner does not know whether the heating is evenly distributed, however, as the floor becomes too hot to walk on during firing.

Temperatures

"High temperature" is almost meaningless unless you know the "normal temperature" that is implied in the comparison. "High-temperature" tumble dryers and foil tapes may handle temperatures up to around 300°F; whereas "high-temperature" concrete might be intended for 752°F (400°C) or so. "Refractory" refers to materials intended for situations above 1000°F (540°C). "Hot lava" or any red-hot object might be 2000°F (1000°C). Incandescent light bulb filaments routinely reach 4500 to 5400°F (2500 to 3000 C, higher for halogen and theatrical lamps), similar to the surface of the sun.

At temperatures above the intended maximum, most materials either melt (metals and plastics), burn (synthetics, natural fibers, some minerals), or lose structural integrity due to other chemical changes like extreme dehydration. Many materials begin to undergo permanent changes well below their destruction temperature, so it's best to allow a margin of safety.

Some Useful Reference Temperatures
Human Comfort
- "Room temperature" is officially set at 68–70°F (20°C); individual preferences range from about 62°F (15°C) up to 85°F (30°C).
- "Lukewarm" or "bathwater" temperatures are roughly similar to body heat, 100°F (40°C).
- "Hot to the touch" means different things to different people, but for many people the hottest touchable surface is 120–135°F.

Maximum legal heat for a sauna in the US is 194°F (90°C); though the original traditional saunas might run closer to 140°F (60°C) on occasion.

Masonry heaters generally run at sauna-like surface temperatures: 120–250°F (50–120°C), depending on design. Some designs include hotter surfaces or ovens for cooking, baking, and radiant heating across greater distances. (Source: David Lyle, *The Book of Masonry Stoves*, p. 118). Many traditional designs include a bench or bed for occupants to enjoy direct body comfort, as well as a hotter vertical surface for distributing radiant heat into nearby rooms.

Radiant woodstove surface temperatures generally average 330–420°F (170–220°C), with maximum surface temperatures up to 780°F (420°C). Stovepipe temperatures may be similar, but can max out hotter (up to 1435°F [780°C] for stovepipe inner surfaces). (Source: *Wood Heating Safety Research: An Update,* Richard Peacock, Center for Fire Research, National Bureau of Standards, Gaithersburg Maryland 20899, manuscript dated 1987.)

Wood Heat

- Touchably warm, safe temperatures for wood walls and structures are 120–150°F (50 to 66°C). Woodstove testing protocols find the safe distance by measuring how far away walls must be so that they do not exceed 85°F above ambient temperature during the hottest phase of the fire (so, if the ambient temperature is 70°F, the clearance would be considered inadequate if the walls got above 155°F).
- Severe drying/ potential charring occurs at 200°F (95°C). Long-term damage and increased flammability are known to occur in wood heated to these temperatures on a regular basis.
- Charring/breakdown (gases are released, which may ignite): 230–450°F (110–230°C)
- Auto ignition: charcoal/dry pine/dry oak: 660/800/900°F (350/430/480°C)
- Clean burn begins (no smoke, little or no CO): 1100–2000°F (600–1000°C)
- Creosote/chimney fire*: 2000–3000°F (1100–1650°C)
- Theoretical maximum (adiabatic flame) for wood and most natural fuels: 3600°F (1980°C)
- Actual temperatures measured in open flames**:
  - continuous region 1650°F (900°C, small flames) to 2190°F (1200°C, larger fires)
  - flame tips 600 to 930 °F (320 to 500°C, varies with fuel)
  - candle flame peak temperatures: 2550°F (1400°C)

(Sources: *University of Michigan Extension Bulletin E-1388, "The Creosote Problem."
**"Temperature in flames and fires," by Dr. Vytens Babrauskas, *Fire Science and Technology Inc.* www.doctorfire.com/flametmp.html, visited 11/29/2012) (UL-listed appliances in the USA are tested to 115°F above ambient room temperature, or a maximum of about 185°F using a firebrand test of super-hot-burning kiln-dried softwood strips — Ken Rajesky, "Wood Stove Wall Clearances — Primer," http://www.hearth.com/econtent/index.php/articles/stove_wall_clear, 11/27/2012)

A Rocket Mass Heater's Typical Surface Temperatures (Varies with Design and Site)

- Flame path (sides of burn tunnel and heat riser): 900°F to 2500°F depending on design
- Barrel surface: similar to woodstoves — 200°F to 800°F
- Masonry surfaces: combustion unit similar to masonry heaters (up to 200°F); heat-exchange mass surface generally cooler (up to 120°F).

Maximum Temperature Limits for Relevant Materials

Coverings:

- Synthetics: (polystyrene/polyurethane/mylar): 165/250/300°F (75/120/150°C; for example, insulation foam, cushion foam, "high temperature" foil tape)
- Fabrics:
  - High iron: linen, cotton, natural canvas: 390°F (200°C)
  - Medium iron: wool, silk: 300°F (150°C)
  - Low iron: polyamide, acrylic, many synthetics: 230°F (110°C)
- Metals: melting points (catastrophic fail temperatures)
  - Aluminum: 1220°F (660°C)
  - Brass (yellow/red): 1660–1880°F (905–1025°C)
  - Copper: 1980°F (1080°C)
  - Iron (ductile/cast/wrought): 2060–2900°F (1150–1590°C)
  - Nickel: 2650°F (1450°C)
  - Steel: 2600–2800°F (1430–1540°C)
  - Stainless steel: 2750°F (1510°C)

Masonry:

- Portland cement, concrete (also lime mortars, paints, plasters):
  - Steam spalling from trapped water: 220°F (100°C)
  - Degradation/weakening of concrete from 750°F (400°C)
  - Structural failures with many aggregates: 1200°F (650°C)
  - High-temperature concrete (amended Portland cement) can operate at the high end of this range, but is still not suitable for the firebox of a clean-burning fire.
- Clay building brick: 2000°F (1000°C)
- Cordierite refractory (crystalline magnesium alumino-silicate): 2280°F (1250°C)
- Kiln brick (soft brick, insulative brick): 2300–2400–2600°F (1260–1320–1430°C, or as rated)
- Firebrick/silica firebrick: 2850–3000°F (1580–1650°C)
- Alumina silicate refractory: up to 2370–3090°F (1300–1700°C)
- Refractory cements: products are available rated for 2100, 2400, 2800, 3100°F

Mineral Insulation:

- Fiberglass (insulation, gaskets): 1000°F (540°C)
- Mineral wool/Rock wool: 1800°F (1000°C)
- Perlite: 2300°F (1,260°C)
- Alumina silicate refractory: up to 2370°–3090°F (1300°–1700°C)
- Other refractory insulation: see manufacturer's rating for each material — many options are available for 2100, 2400, 2800, some up to 3100; a few go up to 3500°F

(Sources: EngineeringToolbox.com, various manufacturers' product specs.)

Wood-Fired Designs for Other Goals

A rocket mass heater is an effective design for heating people in temperate to cold climates. Thermal mass stabilizes temperatures over time, so this type of heater is best for spaces that are used roughly half the time or more. (Four hours per day may be a practical minimum; if you are not using the space at least four waking hours a day, it's not very convenient to operate the fire, and almost all the stored heat would be wasted during unoccupied hours. For spaces used only to sleep in, consider a heated bed with warm covers rather than full-space heating, to minimize firing time. For seldom-used spaces, many owners use lower-mass designs that offer quicker heat-up and cool-down.

The rocket mass heater also operates completely off-grid, with a human operator and natural draft instead of mechanical controls.

Our design goals:
- fuel efficient
- clean
- off-grid ready = passive draft
- low embodied energy
- renewable fuels
- owner-operated and -maintained

A mass storage heater of any kind may not be the best tool for other purposes, since it takes time and fuel for the thermal mass to come up to operating temperatures. For other goals, consider:

- Cooking stoves: responsive, low-fuel, short-fire cycles — look up "rocket stoves," "the Good Stove" + India, solar cookers, solar ovens. There are some good designs at www.aprovecho.org
- Seldom-used or briefly-used spaces: parlors, guest rooms, ski cabins, churches. Consider passive solar design if possible; boost heat on demand with space heaters such as efficient Rumford fireplaces, small woodstoves, soapstone stoves, heat lamps, or even personal people-warmers (brasiers, beanbags, baked potatoes). See www.chimneysweeponline.com for loads of detail on wood-fired space heaters, or our booklet "Shelter" (E&E Wisner, www.ErnieAndErica.info/shop) for alternatives.
- Fireplace design: Rumford fireplaces are far more effective heaters, and more beautiful, than a modern or pioneer-style box fireplace. Look for tapered throats, throat or chimney shut-off dampers, and optional glass or screens. See www.Rumford.com.
- Furnaces and boilers: Heaters designed to be hidden in the basement, outdoors, or a utility closet, and transfer heat through air or fluids for "central" heating such as steam radiators, radiant floors, and forced-air heating. Less fuel efficient because of heat losses in transmission, and a tendency to overheat unused areas (like spare rooms or the ceiling). There's no denying the convenience of automation for institutions or large, sparsely occupied buildings like production greenhouses. See www.heaterhelp.org.
- Classical masonry heaters: All-masonry heaters developed for traditional local needs: Mandarins' k'ang stoves, Korean ondol, Afghan tawakhenah, Turkish baths, Roman hypocausts, central European tile stoves (kachelofen), Slavic, Nordic designs both artistic and peasant vernacular. References: http://mha-net.org, David Lyle's *Book of Masonry Stoves*.

Appendix 6:
Supplemental Practice Activities

Most people learn to build earthen masonry in a hands-on workshop, by serving on someone else's building crew, or as a trade apprentice. Physical skills are best learned by physical practice under expert supervision. Any complex process gets easier with practice.

When we teach rocket mass heaters in person, we generally include separate hands-on activities designed to build skills. Not all of these skills are necessary to master before building a heater, but they can be very helpful. Here, we are offering a selection of activities that may be useful warm-ups for the skills described in this book.

Please note that we provide a separate set of teacher's notes for qualified workshop leaders, and are not suggesting that readers of this book should go out and begin teaching with this guide. A lot of people who have technical expertise, or have built one or two heaters, become overconfident in their own understanding. Please don't leap out and start teaching others too soon — often it's your fourth or sixth heater that teaches you

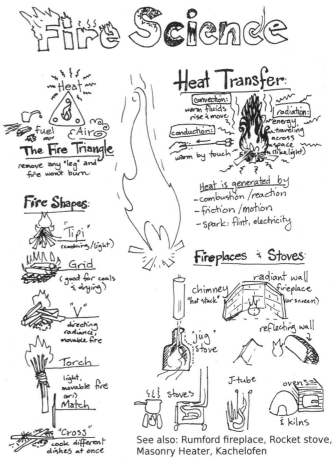

a big lesson. An inexperienced teacher can put others at risk both in the workshop, and over the life of a heater. If you demonstrate something and it doesn't work, it can give people a bad impression of these heaters, and may hurt the prospects for introducing working versions of these heaters in your area. The authors are available for project design consultations and peer mentoring.

Fire Science Practice

A bonfire after dark demonstrates many basic concepts. Food and social relaxation are nice bonuses.

Fire triangle: Fuel, air, heat

Try putting out a fire by depriving it of one of these three things.

Can you imagine situations in which all three elements might be present, yet fire dies or doesn't start? Fire is a chain reaction, not an object.

Fire Practice

Safe Hearth

Start by digging out a safe outdoor hearth (6′ diameter or larger space with leveled, bare mineral soils — many gravel driveways work well, asphalt doesn't). In this space you have plenty of room to move around your projects and drop hot items safely.

Have two extinguishers within a moment's reach, such as (at least) 5 gallons of water, a charged hose, a shovel and sand, or a Type A or ABC fire extinguisher.

Check local fire regulations and weather conditions before burning. Rain doesn't stop us if we have enough dry fuel, but we will cancel any outdoor fire demonstration if there are high winds or a burn ban.

Brick Stacking

Build several brick hearths. Try a flat hearth, a fireback (straight wall), a box-like fireplace, and/or a simplified Rumford (angled) fireplace. Build a similar fire in each one (the V-fire from our booklet *Art of Fire* is a good, versatile option).

Walk around each fire with your hands out; try to find the shape of the fire's warmth (the places where it is comfortable, not too hot for safety). Observe the differences in smoke, fire efficiency, and radiant heat coming off the brick walls, around the fire, above the fire.

If time allows, run the fireplaces about equally for a few hours — enough to warm the bricks through. Compare their warmth shapes again.

Tea Stove Challenge

Place a pot of water of similar size and shape in different relationships to the fire. Try to build a stove using bricks, dirt, and the firewood itself to bring one liter of water to a full boil quickly. (Our favorite stoves average about 6 minutes.)

Chimney Effect

With the fire relatively low, or struggling to engage new fuel, hold a metal stovepipe section over the fire (using oven mitts or large tongs). Try the stovepipe vertically, at a 30 degree angle, at a 45, horizontal. Listen for draft. Compare the fire's action near the chimney and away from its effects. At what angle does the chimney stop drawing? How long does it take a cold chimney to start drawing?

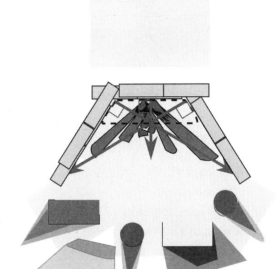

Five-minute Fire

The 5-minute limit is designed to test your ability to recognize the natural fuels around you. It can be fun to do it with larger groups as a way to share relevant knowledge. (If some people in the group are more outdoorsy than others, try assigning them in mixed teams, so that nobody has to feel stupid.)

In any given setting, you will see certain people make a beeline for a particular type of tree, bush, or lichen, or even for the woodshed or recycling bin, while others wander around or base their guesses on experience in a completely different climate.

Take five minutes to gather dry fuel and tinder from the local environment, and lay out a fire in a safe firepit.

After the 5 minutes are up, stop. (Gather the group together, if applicable.) Try to light your fire(s) with one match (or a shared lighter). Success is easy to spot: fires that light right up with one match, a second or two of lighter time, or even a tiny coal from another fire.

Can the group light all the fires from only one match, by passing the fire along?

Modify any unsuccessful fires; see how many types of fuel or layouts you can get to be successful when properly combined and arranged.

Some Notes on Fuels

a piece of wood from your woodpile into 1˝-wide pieces, weigh them immediately, then dry them in an oven for about two days (low baking temperatures below 250 °F, using fan if available). Weigh them again to find the difference.

There are lots of other ways to evaluate fuel moisture content, for example by watching cracks in the logs, by feel, by how it is to split, or with a moisture meter. The weight method is a good way to double-check your other methods; you don't need to do it very often.

Look for fuels with good exposure to air, yet sheltered from weather and ground damp. Living materials like green moss and lichen tend to be very moist in most climates. Compare cordwood that has been stored differently, for different lengths of time, or that was split before storage vs. split just prior to use. How does the fuel behave while splitting — does it spring apart, or shred and cling?

For an accurate fuel moisture percentage, compare the fuel's weight before and after artificial drying. Split

Skills Practice

Skills to Learn Before/During First Installation

- Masonry: Prepare, test, and use earthen building materials like clay slip, clay-stabilized perlite, mortar, cob, and plasters. (Practice projects might include a scale model of a stove or house, like a masonry dollhouse, or J-style outdoor cookstove.) Work to a consistent finish dimension (fieldstone, brick, plaster), including appropriate pre-finish levels for plasters or tiles.

- Stack bricks level and plumb, with and without mortar, and with a thin-set or clay slip. You can re-use bricks with clay mortars, just soak off the clay afterward (or hose it off). If using refractory mortars, practice with a few sacrificial bricks to learn water content, working times, and dimensional accuracy. Practice dry-laying (without mortar) the correct dimensions of firebox, proportions of heat riser to burn tunnel and feed. (Scale mock-ups; feel with fingers to make sure feed is completely flush and smooth.)

The "Thermosiphon" Effect

Many people find it puzzling, amazing, or seemingly impossible that we can make a fire burn sideways and upside down. The following exercises help you to explore this effect first hand.

With bricks and a ready supply of dry kindling, build a mock-up rocket heater core. Build one model that fits the firebox dimensions given in Chapter 4. Think over any questions you might have about proportions, heights, limits to length or size, or the effects of other changes in design. Can these ideas be tested easily?

Build two side-by-side mock-ups to test one of these ideas, like what happens with a larger heat riser. Or a longer burn tunnel. Exaggerate the difference to compare the results.

You want to have a *control* example. So test only one difference for each pair of mock-ups; it can be difficult to interpret the results when two or more changes are made simultaneously. For a fair comparison, both examples should be fired using similar fuels and preheated for similar amounts of time — share hot bricks equally between the two models, for example, if they are already warm from other activities.

A brick-and-dirt mock-up comparison will not necessarily give accurate performance results for a well-built installation. For example, metal fireboxes or noninsulated heat risers may work OK for a 20-minute outdoor test, but are almost guaranteed to fail within their first year of indoor use. (A noninsulated heat riser generally stops drafting about 40 minutes into the fire when hooked up to a full-mass heater.)

This mock-up process is just to get practice stacking brick to internal dimensions, and to understand possible effects of proposed changes by exaggerating them while holding other factors roughly equal.

Clean burn and steady draft are the desired results (no smoke out of either end). Watch for general effects that make the fire burn cleaner or smokier, draw faster or more sluggishly, or stop the stove from working as intended. Try comparing different sizes, heights, leaky vs. mortar-sealed, insulation vs. thermal mass, size of firewood, etc.

- Pipe and ducting: Use crimpers, tinsnips, grinder or other cutting equipment, with appropriate safety gear and spark-safe settings. (Consider making a coffee-can barbecue starter to practice with tools — cut it from a big steel can; crimp it just for fun; tap or pre-drill holes for a handle.) With clean hands, practice setting adjustable pipe elbows to any desired measurement: 90°, 60°, 45°, 30°.
- Metal and masonry skills for manifold/barrel seals: cutting, fitting, placing insulation, expansion joints, cobbing, plastering. (A pocket rocket to burn paint off the metal barrel(s) is good metal-cutting practice. If you find you enjoy working with brick far more than metal, and can get airtight mortar joins consistently, consider building a brick manifold instead.)
- Level and plumb: Check vertical alignment of the chimney pipe and barrel with a plumb line or level. (Practice bringing the barrel to level using shims under it. Level a bed of sand for your firepit, remembering to check diagonals. Check your level by swapping ends — is your

level actually level? Watch out for mud and grit stuck to the bottom.)

Personal Training

In learning new physical skills, give yourself the best conditions for success. Food, water, adequate rest and stretching, good work conditions, and a good chunk of focused time. If you can work directly with an experienced trainer, or with good video documentation, you can avoid a lot of wasted effort.

The point of practicing is to develop habits you can rely on when you get to the more complicated parts. Begin as you mean to go on: practice safe body movements (lift with legs not back, use arms more than wrists); develop good tool care and site organization; always follow tool manuals and safe positioning. Start with slow and correct movements, and smaller loads, and work up: a wheelbarrow full of masonry is a LOT of weight!

Like riding a bicycle, most physical skills are tough to learn but easy to maintain once learned. Learning movements and postures correctly is as important as knowing the theory. Once the basic skills become automatic, complex tasks become much easier.

Sports trainers often drill only one new skill per 4-hour training period, to let players assimilate the lesson into long-term physical memory. (Other, familiar skills can still be practiced during this period without interrupting the process.) Make sure to end on a correct performance of the new skill. You can also rehearse the correct process in your mind between practice sessions.

Barrel Decoration

Many readers wonder if we can make the barrel more beautiful or substitute a better-looking bell. Of course there are many ways to change the barrel, but some changes work better than others. We know that certain changes, like wrapping insulation around the barrel, will cause major problems with the heater's draft performance. We want attractive options with few or no side effects to hurt performance.

The following pages show the Barrel Deco Design Challenge, a course handout developed for a metal-working class. These pages show some examples and ideas for barrel decoration, and go over important functional and safety requirements to keep the heater working as intended. Readers

Barrel Decorating: The metal cylinder works.

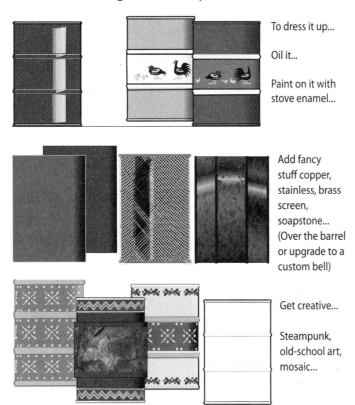

To dress it up…

Oil it…

Paint on it with stove enamel…

Add fancy stuff copper, stainless, brass screen, soapstone… (Over the barrel or upgrade to a custom bell)

Get creative…

Steampunk, old-school art, mosaic…

who like metal work may also be interested in making custom components or accessories, such as fireplace tools.

Barrel dimensions:

55-gallon drums (200 L) are a volume standard, but vary slightly in height/diameter. Many foodgrade barrels are 23⅝" diameter and 34½" tall (at outside top rim).

Barrel operating temperatures for the top third of the barrel reach 400–800°F; the bottom third rarely reaches 300 °F. Center top surface is convenient for simmering and frying; mostly it runs well under 1000°F. Occasionally shows dull red spot or ring for some designs (1200°F?).

Rocket Mass Heater Barrel Design

The barrel of a rocket mass heater serves to seal in and redirect exhaust gases, and to shed heat. A weathered black steel cylinder radiates heat into a room very effectively and evenly. This heat output is also a performance requirement: the heat lost by the barrel helps maintain downward exhaust flow for proper draft.

If we want a betterlooking object to serve these functions, what might we use?

One option is to substitute a better barrel:

- 55-gallon shipping barrels (food or fuel grade)
- 35-gallon or 120lb grease cans (Jiffy Lube trash cans) for smaller systems; selection of styles
- Hot water heater tanks (with insulation and innards removed)
- custom-welded stainless or mild steel cylinders
- large square steel containers

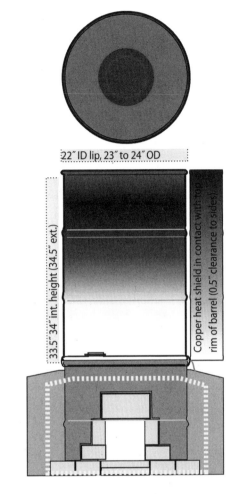

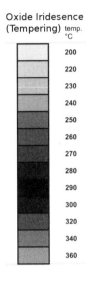

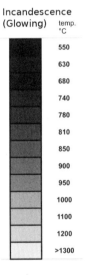

- big metal garbage cans with galvanization removed. All must be stripped of paint, galvanization, and/or insulation before installing. We have no reports whether the glass enamel coating remains intact on the inside of water heaters.

To beautify existing barrels, we have tried:

What worked:
- Coating the barrel with earthen plaster or lime plaster up to one third of the surface. Covering the whole barrel can cause overheating and draft problems.
- Painting the barrel with high temperature cooking oil, stove enamels or engine block

paints (rated 1200°F or higher), mineral paints like clay or lime (unrated pigments may change color).

- Pizza stones, soapstone, marble, ceramic tile, earthen plasters/clay-sand grouts directly on top of the barrels: no problems. DO NOT USE SLATE in direct contact with top parts of barrel, as trapped steam may cause spalling or explosion.
- Wrapping or suspending *"a metal heat shield"* around the barrel, with air gap. This works really well and we'd like to try more styles. Standard heat shielding uses a 1″ gap behind the sheet, plus 12″ gaps above and below for natural convection.

What didn't work:

- Insulative materials around the barrel almost always cause major draft problems (perlite, rock wool, structures that trap air without enough vents).
- Wrapping or mounting cement board (tile backer) or gypsum board within 3″ of the barrel did NOT work. These materials offgassed, charred, cracked, and decomposed. Air gaps top and bottom might have helped; but we also would look into higher-temperature board products specifically designed for heat shielding on/near woodstoves or hearths (not just for wall-mounted heat shielding).
- Still working on a "nonmetal" option. Brick "bells" don't draft this way, but true masonry heaters are all brick and tile (see Swedish contraflow, Finnish, and Russian fireplaces).

We've had some luck with 20–24″ terracotta chimney liners as a barrel substitute,

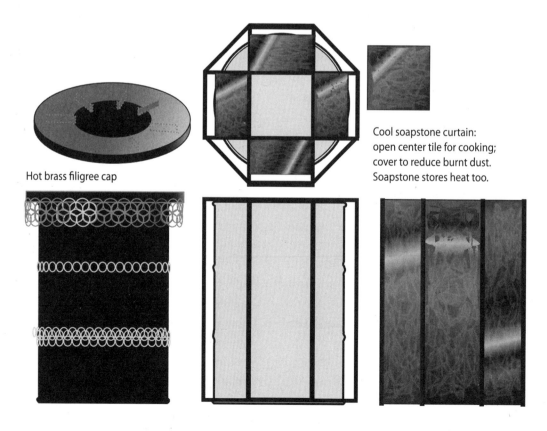

Hot brass filigree cap

Cool soapstone curtain: open center tile for cooking; cover to reduce burnt dust. Soapstone stores heat too.

but have not yet created a good capping slab blast shield for the top. Mortar seals are problematic and prone to cracking over repeated firing cycles; double layers of tile may be needed to cover joints.

We'd be interested in "tile stove" components with a box-like or fin-like heat-shedding structure, but have not yet tried them. Blast shield for top of barrel (on inside) to reduce warping?

My personal favorite design proposal is a metal "spider" that would rest over the top of the barrel and support decorative stuff hanging down by the sides.

A short version could just be a "cap" or "doily". Brass filigree lacework; or turn a fluted garbage can into a faux Grecian column with a faux capital and hi-temp paint faux marbling. (Enough faux faux youx?)

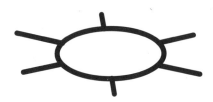

A long version of this "curtain wall" system could extend nearly all the way down the sides (or all the way down except for a hidden air gap in back). This would create a support framework for any heat tolerant surface including stone slab, punched tin, tile, filigree work, enamel work, or other attractive materials.

Note that both my examples may still not go with that particular sofa.

Design Goal:
Barrel or bell that is elegant enough for a fine-homes magazine, safe, and functional.

Design Constraints:
The barrel or replacement must stay sealed, tolerate high temperatures, and shed heat.

1) Air seals: Avoid piercing the barrel. Welds or bolts may cause stress cracking; consider assymetrical expansion with hot/cool cycling. Use symmetry, expansion joints, slots vs. holes.

2) Hot: Think stone, metal, ceramic tile, mineral grouts (clay or lime), paints rated for 1200^{+}°F. No acrylic, latex, Portland cement, or unrated goop (even if it says "high heat").

3) Shed heat: Convection (air gaps), radiant heat, conduction. No insulation or trapped air.

4) Maintenance access: Allow removal either of the entire barrel (access at bottom 4″), or the barrel's lid from the top.

5) Home safety: Make designs robust enough for "normal" use: toddlers, parties, spills, pets, step ladders, and winter storm power failures.

Bonus Features:
- Cooking? (center top surface for simmering, top as griddle...)
- Oven? (insulated dome over top?)
- Heat shielding for wall sides / back? (reduce inconvenient clearances to combustibles)
- Extra exposed barrel in front/on room side?
- Adjustable room heat? Doors/fans to control radiant and convective heat delivery? Directional heat settings? (Use any electricity as a bonus not a "fix" be off-grid ready.)
- Safe drying racks/hooks? Mittens, kindling ...

Here are some blank barrels, for drawing out your design ideas:

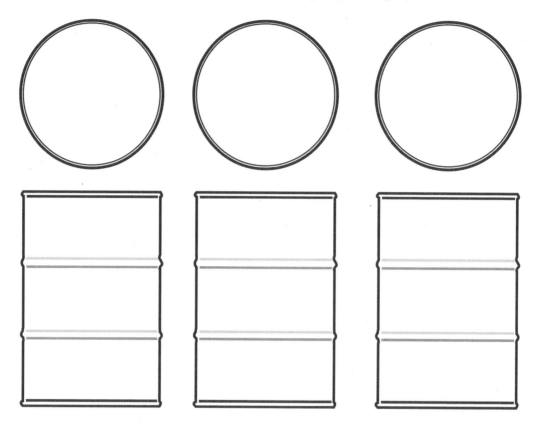

- Low maintenance? Cleans with ordinary tools/vacuumtips? Durable? (Problems unlikely, easily detected, fixable?)
- Matches an existing room: tile backsplash, brick hearth, copper kettles....
- Allows redecorating (color changes, resurfacing, uncut standard-sized tiles or bricks)
- Cosy fireside comforts... watch the fire? Safe/comfy to lean on? Reheat beanbags or hot water bottles? Cup holders?

Rocket Accessory Projects for Metal Workers

Fireplace Tools: S-shovel

For our J-shaped firebox, we could use an ash scoop with a specially shaped handle. (Hands or a tin can works fine, but a handle seems more elegant). It could work like a soup ladle or minibackhoe, to drag and then scoop up ash.

Narrow tongs (like for barbecue grills, but with longer fireproof ends and shorter combustible handles) are very useful. Like giant tweezers. Gentle spring is excellent.

Super Bonus: Manifold

(alternative metal crafting project):

Use a cut off barrel, or improve on it:

- 16" to 18" height
- ports in back/side to fit 8" ID stovepipe (two interchangeable ports? exhaust pipe, cleanout)

- rectangular or stepped cutout (bent flanges?) to fit over brick firebox, insulation, and mortar at least 10" tall x 13" wide, see above.
- improved clamping or gasket seating fitting: we use a band clamp and sticky-backed gasket.

Make a groove so barrel rests on the gasket? Sturdy, airtight interior flange to resist side impacts, yet remain removable?

Visit our web page to share examples and design ideas: http://www.ErnieAndErica.info

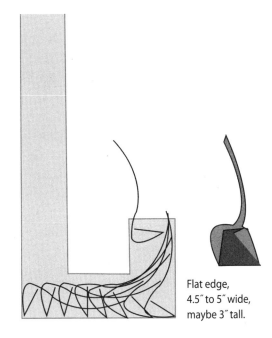

Flat edge, 4.5" to 5" wide, maybe 3" tall.

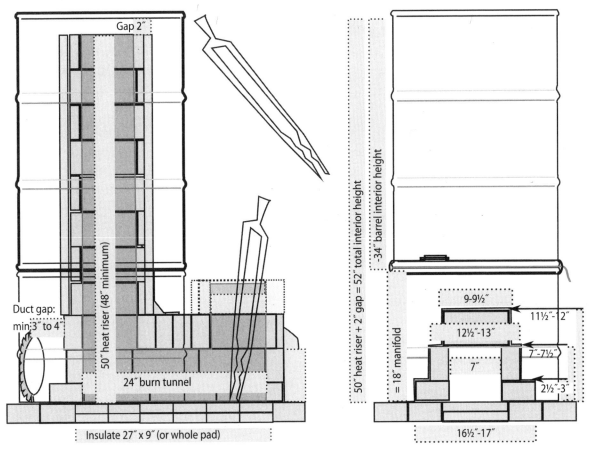

Online Resources

Cob Cottage Company in Oregon (www.cobcottage.com)

Graeme North in NZ (www.ecodesign.co.nz)

Mike Wye in the UK (www.mikewye.co.uk)

Sweep's Library, http://www.Online Chimney Sweep.com, www.permies.com/forums

YouTube videos — look for finished projects and 1-year-updates

Paul Wheaton, www.richsoil.com/rocket-stove-mass-heater.jsp, includes videos

Ernie and Erica Wisner: plans, books, videos, articles, project photo albums, etc.: www.ErnieAndErica.info, www.ErnieAndErica.blogspot.com

Fireside References (books with other good info or projects)

Dunn, Kevin M. *Caveman Chemistry*, 2003, Universal Publishers, www.CavemanChemistry.com

Evans, Ianto, Linda Smiley, Michael Smith, Deanne Bednar. *The Hand-Sculpted House*, 2002, Chelsea Green Publishing Company, Burlington VT.

Evans, Ianto and Leslie Jackson. *Rocket Mass Heaters*, 2006 and 2014, Cob Cottage Company, www.rocketstoves.com.

Lyle, David. *The Book of Masonry Stoves*, 1998, Chelsea Green Publishing, Burlington VT.

Orton, Vrest. *The Forgotten Art of Building a Good Fireplace.* Yankee Magazine, 1969

Ritter, Erica. *The Ultimate Girls' Guide to Science,* 1993, Beyond Words Publishing, Hillsboro OR.

Wisner, E.K. (Wisner, Ernie and Erica) *The Art of Fire,* 2013. Self-published, http://www.ErnieAndErica.info

Young, Jon, Ellen Haas, and Evan McGown. *Coyote's Guide to Connecting with Nature,* 2010, Owlink Media Corporation, Shelton WA.

Index

Page numbers in *italics* indicate illustrations.

A

aggregates, for cob, 186
air supply. *See* outside air intake
air-to-fuel ratio, 246
aliz paint, 196
Alliance of Masonry Heater and Oven Professionals, 153
Annex 6″ example, *40*, 41–42, *42*
area calculations, 249–250
The Art of Fire (Wisner), 21, 241
as-built drawings, 106
ash
  buildup, 111, *111*
  cleaning, 119–121, 122, 148, 245–246
ASTM (American Society for Testing and Materials), 140
ASTM 2778, 152
ASTM E-1602, 152, 156–160
Audel Masons and Builders Library, 189

B

Barker, Tim, 213, 214
barrel. *See* bell
basements, 169
base pad, example installation, 75–76, *76*
batch box example, 221–223
bell, *4*
  clearances, 64–65
  decoration, 133, 262–266
  design errors, 174–175
  example construction, 82–84
  finishing, 90, 172
  options, 62
  removing paint, 83
  requirements, 8–9
  shielding, 156–157
bench. *See* heat-exchange bench
bentonite clays, 185
Bobcat machines, 60
boilers, 103, 164–165
Bonny 8″ example, *43*, 44–45, *44, 45*
The Book of Masonry Stoves (Lyle), 28, 138, 180
brick facades, 196
bricks. *See* firebrick
bridge
  brick inspection, 124
  example construction, 78–79
  requirements, 7–8
BS-EN standards, 153
BTUs (British thermal units), definition, 226
building codes
  for chimneys, 70
  clearances, 64, 65

design change requirements, 153–154
    for masonry heaters, 140–141
    in Portland, Oregon, 199–203
    for rocket mass heaters, 152–161
burnished finish, 195
burn tunnel, *4*
    dimensions based on brick size, 250
    inspection, 124
    requirements, 6–7, 145
bypass dampers, 116–117, *116*

C

Cabin 8" example
    design, *46*, 47
    heat delivery calculations, 149
    heat storage, 152
calculations
    chimney draft, 248
    heat delivery, 149
    heat loss, 225–233
    RMH dimensions, 248–252
carbon monoxide exposure, 135, 139
casing, 14–15, 90–93
    See also heat exchanger; thermal mass
cement mixers, 60
cements, 181–183, 193–194
ceramic-fiber blanket, 81
channel. *See* heat-exchange channel
chimneys
    clearances, 66
    code requirements, 159
    design, 29, 146–147, 171
    design for efficiency, 236–239
    draft calculation, 248
    example construction, 70
    laminar flow, 247
    location of exit, *4*
    maintenance, 242–243
    orientation of, 70–71, *70, 71*
    priming, 116
    requirements, 15–16
    through-roof installation, 72–74, *72, 73*, 238
    through-wall installation, 205–207, 238–239
    troubleshooting, 130–132, 133–135
    use of existing, 170, 207–208
cladding. *See* casing
clay, 185–186. *See also* earthen masonry
clay paint, 196
clay-sand mortar, recipe, 125, 188–189
clay slip
    brick laying method, 77, 78–79
    preparation, 76–77
    recipe, 187–188
clay-stabilized perlite, 81, 188
cleanouts
    chimney, 16
    clearances, 66
    covers, 94–95
    definition, 105
    in example construction, 86–87
    in heat exchanger, 12
    locations, *4*, 144
    in manifold, 10
cleanup, site, 67
clearances
    code requirements, 156–157
    design changes, 171–172
    for example installation, 64–66, *64*
    walls, 170
coal, 220–221
cob
    application, 194–195
    batch consistency testing, 192
    finishing, 90–91, 195–197
    for heat-exchange benches, 87, 88–89
    as insulation, 81
    materials for, 184–187
    methods, 183–184
    mixing, 60
    recipe, 191
    test bricks, 190–191
The Cobber's Companion (Smith), 183
codes. *See* building codes
cold starts, methods, 115–117
combustion unit
    about, 2, 4
    components and requirements, *4*, 5–10, *105*

definition, 105
design constraints, 142–143
example construction, 75–87, *99*
example construction (6" alternative), *96, 98*
measurements, *75*
role of, 4–5
conduction, 27, 86, 148
construction
   considerations, 29–34
   documentation, 107–109
   errors in design, 150–151
contraflow bell. *See* bell
convection, 26, *26*, 246–247
cooking, 24, 165
cooling degree days (CDD), 227
core, 13–14
   See also heat exchanger; thermal mass
cottage, heater example, *46, 47, 48, 49*
crack repair, 126–127
creosote, 244
cross-sectional area (CSA)
   bell requirements, 9
   consistency in, 5
   design constraints, 142–143
   in heat exchanger, 11–12
   manifold, 9
   pipe diameter and, 106, 251–252

D

dampers, 139
Daybed 6" example, *48, 49*, 152
dead men, 94, *94*, 195
design
   alterations to, 133, 171
   changes by building officials, 153–154
   code-required constraints, 154–156
   common errors, 150–151
   CSA, 142–143
   examples of, 39–58
   lifestyle and, 36–37
   location and, 25, 27–29
   room use, 31–34
   separating components, 170
Dezettel, Louis M., 189

documentation
   access points, 121
   of construction, 107–109
   performance log, 118, 120
   preliminary inspection, 110
downdraft bell. *See* bell
draft
   control of, 147–148
   effect of CSA, 5, 251–252
   testing, 111
   types of, 249
drawings, as-built, 106
ducting. *See* heat-exchange channel
dung, 193

E

earthen masonry
   advantages of, 90, 168–169, 180–182
   finishing, 195–197
   historical use of, 179–180
   maximum temperatures, 254
   refinishing, 128–130
   repairs, 124–127
   test-firing, 89
earthen plasters, recipe, 192–193
egg tempera, 197
8" ID system, combustion unit measurements, 142–143
   See also example construction
Ely 8" example, 214, *215*
emergency shutdown, 113
energy efficiency
   calculations, 33
   improvements to, 31, 233–235
EPA (Environmental Protection Agency), 140, 152, 160
Europe, building standards, 153
Evans, Ianto, 71, 83, 101, 140, 180, 183, 211, 243
example construction
   base pad, 75–76, *76*
   bell and manifold, 64–65, 82–84
   brick layout, *97, 99*
   bridge, 78–79
   chimney, 66, 70

cleanouts, 66, 86–87
clearances, *64*
combustion unit, 75, *75*, 99, *99*
finishing, 90–95
firebox, 65, 76–79, *77, 78*
foundation, 69–70, *69*
heat-exchange bench, 65–66, 88–89
heat-exchange channel, 85–86, *85*
heat riser, 78, *78*, 79–80, *79, 96*
insulation, 80–81, *80*
location, 67
materials, 60, 62–63
6″ alternative, 96, *96, 98*
specifications, 59–60, 63
test-firing, 89–90
thermal mass, 87–89
examples
    Annex 6″, *40*, 41–42, *42*
    batch box, 221–223
    Bonny 8″, *43*, 44–45, *44, 45*
    Cabin 8″, *46*, 47, 149, 152
    Daybed 6″, *48*, 49, 152
    Ely 8″, 214, *215*
    Greenhouse 8″, *55*, 56–58, *56, 57, 58*
    Mediterranean 6″, *50, 51*, 52
    mosquito heater, 223–224
    Phoenix 8″, *53*, 54
    Zaytuna Farm water heater, 213–214, *213*
exit chimneys. *See* chimneys
expansion joints
    cracks, 127
    insulation as, 148
    for metal manifold, 85
    test-firing, 94–95
    use of, 35

F

fabrics
    for heat-exchange benches, 15
    inspection, 122–124
    maximum temperatures, 254
fans, 117, 147
fiberglass insulation, 81
fibers for cob, 186–187
finishing, of earthen masonry, 195–197

fire
    air-to-fuel ratio, 246
    characteristics of, 241
    chemistry, 243–244
    hazards, 242–243
    minimizing smoke, 244–246
    practice activities, 258–259
firebox
    clearances, 65
    design changes, 167
    design errors, 173
    dimensions based on brick size, 250
    example construction, 76–79, *77, 78, 97*
    installation, 166
    materials for, 169, 182–183
    mock-up, 68
firebrick
    base pad assembly, 75–76, *76*
    burn tunnel dimensions based on size, 250
    firebox assembly, 76–79, *77, 78*
    heat riser assembly, 79–80, *79*
    mortar and, 76
    replacement in bridge, 7–8
fire starting
    cold start methods, 115–117
    routine starts, 109–113, *111*
    troubleshooting, 130–132
firewood. *See* wood
floors
    building support for heater, 69–70, *69*
    choosing location, 169–170
    wood, 34, 44–45, 170
flues, 207
footings, *30*, 34, 69
fossil fuels, 220–221
foundations, 29–31, *30*, 69–70
freeze protection, 29
fuel, 117–119, 218–222
    See also wood
fuel feed, *4*
    horizontal feed, 173
    measurements, 145
    requirements, 6
furnaces, 103, 164–165

G

gas flow, cross-sectional area, 144–145
Greenhouse 8″ example, *55*, 56–58, *56, 57, 58*
guest shed, heater example, *48*, 49

H

The Hand-Sculpted House (Evans, Smith, Smiley), 71, 180, 183
hardeners, 193–194
hearth, 65, 157–158, 242
heat delivery calculations, 149
heaters
    choices in, 36–37
    definition, 103
    energy calculations, 33
    options, *23*, 233–235
    portable, 214, *215*, 216–218
    See also rocket mass heaters
heat-exchange bench
    clearances, 65–66
    example construction, 65–66, 88–89
    surfaces, 87–88
    See also heat exchanger; thermal mass
heat-exchange channel
    about, *13*, 106
    code requirements, 158
    design, 143, 146–147, *147*
    diameter effect on CSA, 251–252
    in example construction, 85–86, *85*
    requirements, 11–12
    texture in, 144
heat exchanger
    clearances, *69*
    components, *3*, 11–15, *13, 105*
    definition, 105
    design errors, 175–177
    function, 10–11
    gap measurements, 250–251
    mock-up, 68
    See also heat-exchange channel; thermal mass
heating degree days (HDD), 227
heating methods, *23*, 233–235
heat loss, calculations, 225–233
heat production, 163–164

heat riser, *4*
    design errors, 174
    example construction, 78–80, *78, 79, 96*
    gap measurements, 250
    insulation, 80–81, 148, 173–174
    requirements, 8, 145
heat storage capacity, 151–152
heat transfer, 26–27
Holmes, Lasse, 221
home energy checklist, 31
homes, energy efficient design, 233–236
horizontal exit chimneys, 205–207, 238–239
hot water heating, 24, 213–214
human comfort temperatures, 252–253

I

infiltration, heat loss calculations, 230–231
inspections
    bridge brick, 124
    masonry surfaces, 122
    routine, 109–111
    upholstery, 122–124
insulating brick, for heat riser, 81
insulation
    about, 11
    in heat riser, 8, 80–81, *80*, 148
    materials for, 86, 148
    maximum temperatures, 254
    R-value estimates, 228–230
International Building Code B1002, 152
International Residential Code (IRC) R1002, 152, 155–156, 208

J

Jackson, Leslie, 83, 101, 211, 243
J-style fireboxes
    components, *5*
    requirements, 5–6
    as thermosiphon, 247–248

K

kilowatt hours, definition, 226

L

laminar flow, 247

lava rock, 81
Lawton, Geoff, 213
lifestyle, effect on heater choice, 36–37
lime paint, 197
lining material
    for firebox, 5
    for heat riser, 8
location
    basements, 169
    choosing location, 25, 27–29
    for example installation, 66–67
Ludwig, Art, 234
Lyle, David, 28, 138, 180, 181

M

maintenance
    ash cleanout, 119–121
    bridge, 124
    inspections, 122–124
    masonry repairs, 124–127
    refinishing, 128–130
make-up air. *See* outside air intake
manifold
    design errors, 175
    example construction, 81–86, *82*
    gap measurements, 250–251
    requirements, 9–10
    using barrel, 266–267
masonry
    maximum temperatures, 254
    skills practice, 260–261
masonry chimneys, existing, 170, 207–208
masonry heaters
    building codes, 140–141
    designs, 138–139
    temperatures, 252–253
    use of earthen masonry, 179–180
masonry surfaces, inspection, 120–121
materials
    for example construction, 60, 62–63
    insulation properties, 86
    types of, 35
Mediterranean 6″ example, *50, 51*, 52
metakaolin, 186
metals, maximum temperatures, 254

metal trim, 94–95
metric dimensions
    150 mm example, *98*
    200 mm example, *100*
milk paint, 196–197
mock-ups, 68
Mongolia, 220
mortar, 76–77, 189
mosaic work, for thermal mass finishing, 92–93
mosquito heater, 223–224
multi-story house, heater example, *43, 44–45, 44, 45*
Myrtle Library, Coquille, Oregon, 183

N

North American Masonry Heater Association, 138, 153

O

occasional-use spaces, 22
150 mm (6″) example dimensions, *98*
operation
    cautions, 101–102
    fuel addition, *112*
    routine, 109–113, *111*
    smoke issues, 130–132
Oregon Cob method, 183–184
outside air intake
    code requirements, 159–160
    design requirements, 12
    heat loss calculations, 230–231
    problems of, 34, 173, 208–209
    prototype, *210–211*

P

paints, 128, 196–197
P-channel metal air intake, 124
pellet fuel, 218–220
performance log, 119, 120
perlite, 81
perlite, clay-stabilized, 81, 188
permits. *See* building codes
petroleum fuel, 220–221
Phoenix 8″ example, *53*, 54

pipes. *See* heat-exchange channel
placement of heater, 25, 27–29
plasters
   finishing, 90–91, 128–130
   finishing cob, 195
   recipe, 192–193
pocket rockets, 83
portable heaters, 214, *215*, 216–218
portable structures, 24
Portland, Oregon, building code, 199–203
Portland cement, 35, 182
pumice, 81

R
radiant heat, 26–27, *26*, 28
refractory cements, 35, 181–183, 254
refractory insulation, installation, 80–81
refractory masonry, test-firing, 89
refractory materials, definition, 180
refractory mortars, brick laying method, 77–79
removal of heater, 132
rental apartment, heater example, *40*, 41–42, *42*
resources, 135–136, 197–198, 268
Rocket Mass Heaters (Evans and Jackson), 83, 101, 211, 243
rocket mass heaters (RMHs)
   about, 1–2
   BTU production, 32
   building codes, 152–161
   calculations, 248–252
   design assumptions, 103–105
   design considerations, 16–17
   design goals, 254–255
   efficiency of, 163–164
   location of, 27–29
   parts of, 2, *3*, *104*
   pros and cons, 16–24
   temperatures, 253–254
rock-wool insulation, for heat riser, 81
roof, chimney installation, 72–74, *73*
rototillers, 60
rules, types of, 137–138
   See also building codes

S
safety, 164, 171–172
sand, 186
seat test, 34
shutdowns, 113
Simple Shelter (Wisner), 236
sites
   design considerations, 25
   laying out system, 68
   preparation, 67
6″ example
   combustion unit layout, *98*
   dimensions, 96, *96*
size, 165, 221–224
Smiley, Linda, 71, 180, 183
Smith, Michael G., 71, 180, 183
smoke
   cause of, 243
   clean fire production, 244–246
   issues, 130–132, 133–135, 242
S-shovels, 266, *267*
stone, for thermal mass finishing, 91–92
stoves, definition, 103
straw, 186–187
straw-cob, for insulation, 81
structural cob, 190–192
stucco, 197
system size, CSA and, 106

T
temperatures
   of casing, 15
   chimney, 16
   combustion unit, 4–5
   commonly used standards, 252–254
   design and, 146
   heat riser, 8
   safety and, 114
test-firing
   after manifold construction, 85
   with cob, 95
   procedure, 89–90
thermal cob
   definition, 190
   recipe, 125, 191–192

thermal expansion, 127
thermal mass
  about, *13*, 107
  alternatives, 212–218
  design changes, 168, 172
  in example construction, 87–89, *87*
  heat storage, 151–152, 167–168
  portable, 214, *215*
  requirements, 12–14
  test-firing, 89–90
  troubleshooting, 133
  See also heat exchanger
thermosiphons, 5–6, 165–166, 247–248, 261
therms, definition, 226
through-roof chimney installation, 72–74, *72*, *73*, 238
through-wall chimney installation, 205–207, 238–239
tile, for thermal mass finishing, 91, 92–93, 195–196
tiny homes, 235
tongs, 266, *267*
tools, 60, 61, 266, *267*
Torcellini, Rob, 218–220
tractors, 60
trim, 93–95
Truovo, Francesco, 249
turbulence, 247
200 mm (8″) example dimensions, *100*

U

unit masonry, 87–88, 196
upholstery. *See* fabrics
U-trap function, 246

V

van den Berg, Peter, 124, 161, 221–223, 247
vapor barriers, 197

veneers. *See* casing
ventilation, 209, 230
vermiculite, for heat riser, 81
volume calculations, 249–250

W

warm climates, heater example, *50*, 51
water
  heating by RMH, 24, 213–214
  as thermal mass, 212
weather
  chimney design and, 146
  fire starting and, 117
  operation problems and, 133
weight loads of foundations, 30–31
Wheaton, Paul, 216
Wheaton Labs, 210–211
whitewash, 197
wing walls, 158
wood
  burn temperatures, 253
  drying, 21, 242, 244–245
  heat delivery calculations, 149
  moisture content, 260
  recommended, 2, 117–119
  storage, 241
  trim, 195
wood floors
  building on, 34, 170
  in heater example, 44–45
woodstoves, 244, 246, 253
wood trim, 93–94

Z

Zaytuna Farm, 213–214
zone efficiency, 27–28

About the Authors

Erica Wisner is a skilled educator and project coordinator, with over 25 years' experience building teamwork and facilitating hands-on learning. She loves making things from scratch — anything from blueberry scones to the oven they were baked in. Erica is the author of *The Ultimate Girls' Guide to Science* and *The Art of Fire* and she has also created instructional materials and rocket mass heater resources for the Oregon Museum of Science and Industry, The City Repair Project, Handprint Press, and others. Together with her husband Ernie, she has built over 700 super-efficient, clean-burning masonry stoves and led over 50 workshops across North America.

Ernie Wisner is a jack of all trades — his rich and varied background includes fishing and maritime work, firefighting, botany, natural building, hydraulics, and plumbing maintenance. Extreme conditions and high-risk work focused Ernie's attention on survival skills and emergency preparedness.

Author photo © 2009 by Kacy Ritter.

If you have enjoyed *The Rocket Mass Heater Builder's Guide*, you might also enjoy other

BOOKS TO BUILD A NEW SOCIETY

Our books provide positive solutions for people who want to make a difference. We specialize in:

Food & Gardening • Resilience • Sustainable Building
Climate Change • Energy • Health & Wellness • Sustainable Living

Environment & Economy • Progressive Leadership • Community
Educational & Parenting Resources

New Society Publishers

ENVIRONMENTAL BENEFITS STATEMENT

New Society Publishers has chosen to produce this book on recycled paper made with **100% post consumer waste**, processed chlorine free, and old growth free.

For every 5,000 books printed, New Society saves the following resources:[1]

44	Trees
3,992	Pounds of Solid Waste
4,392	Gallons of Water
5,729	Kilowatt Hours of Electricity
7,257	Pounds of Greenhouse Gases
31	Pounds of HAPs, VOCs, and AOX Combined
11	Cubic Yards of Landfill Space

[1] Environmental benefits are calculated based on research done by the Environmental Defense Fund and other members of the Paper Task Force who study the environmental impacts of the paper industry.

For a full list of NSP's titles, please call 1-800-567-6772 *or check out our website at:*
www.newsociety.com